Positive Parenting

Positive Parenting is a series of handbooks primarily written for parents, in a clear, accessible style, giving practical information, sound advice and sources of specialist and general help. Based on the authors' extensive professional and personal experience, they cover a wide range of topics and provide an invaluable source of encouragement and information to all who are involved in child care.

Other books in this series include:

Talking and your child by Clare Shaw – a guide outlining the details of how speech and language develops from birth to age 11 and how parents can help with the process.

Your child from 5–11 by Jennie and Lance Lindon – a guide showing parents how they can help their children through these early years, stressing the contribution a caring family can make to the emotional, physical and intellectual development of the child.

Your child with special needs by Susan Kerr – a guide for the parents of the one-in-five children with special needs, giving families practical advice and emotional support, based on the shared experiences of other parents.

Help your child through school by Jennie and Lance Lindon – a guide which looks at the school years from the perspective of the family, showing how parents can help their children to get the most out of their years at primary school and how to ease the transition into secondary education.

Help your child with reading and writing by Lesley Clark – a guide which describes the stages children go through when learning how to read and write and shows parents how they can help to boost their children's confidence and give them a headstart in language skills.

Help your child with a foreign language by Opal Dunn – a practical guide to show parents how they can re-use the techniques they used to teach their children their own language to teach them to enjoy speaking a foreign language, even if it is a language the parents do not speak well themselves.

What this book is about

This book aims to help parents

- to understand what school maths involves now;
- to feel more confident about their own mathematical thinking;
- to help their child with maths by suggesting activities to do at home or on outings;
- to involve themselves in the maths education of their child's school;
- to understand the National Curriculum;

and above all

- to help their children to develop a positive attitude to maths and to become mathematical thinkers in their own right.

Health warning: the idea of this book is not to make you feel guilty about what you haven't done! It's never too late to get started and doing just one activity a month with your child is better than nothing.

Contents

H CHILD

with

Maths

A PARENTS' HANDBOOK

SUE ATKINSON

Headway · Hodder & Stoughton

Acknowledgements

I enjoyed writing this book. I hope you and your child will enjoy using it. A number of people very helpfully read drafts and made comments. I particularly want to thank Shirley Clarke, Lynne Atkinson and Lynne McClure; and Ann and Martin Bridges for the Mottik drawings. A very special thank you to Joy Dunn for all her help and advice, especially with the 'family maths' chapter, to Steve Hodgkinson for his wonderful pictures, and to Jonathan and Rachel for being such fun when they were growing up.

Sue Atkinson

London, April 1994

The Publishers would like to thank the children of Charlbury Primary School, Oxfordshire, for their help with the covers for this series.

British Library Cataloguing in Publication Data

Atkinson, Sue
Help Your Child with Maths. – (Positive Parenting Series)
I. Title II. Series
510.7

ISBN 0-340-60767-X

First published 1994
Impression number 10 9 8 7 6 5 4 3 2
Year 1999 1998 1997 1996

Typeset by Rowland Phototypesetting Limited, Bury St Edmunds, Suffolk.
Printed in Great Britain for Hodder & Stoughton Educational, a division of Hodder Headline Plc, 338 Euston Road, London NW1 3BH by Cox and Wyman Limited, Reading.

This book is for Clare, Ellie, Lucy, James, Elizabeth, Leo, Sarah and Nicky.

CHAPTER ONE

Maths – what it's all about

Maths for many people is an area of fear and mystery. Some people seem to be naturally good at it, but for others it is difficult and it brings back memories of failure at school. Even if we were quite good at it at school, the maths that our children bring home from school is confusing and much of it is quite new to us.

Maths as it was taught

For many people, maths was taught from the blackboard. The teacher did one or two examples on the board of how to do it, then the children were told to open their textbook where there would be pages and pages of examples to work through! If you couldn't do these, the teacher would do yet another example on the board. There was rarely any explanation of what it was all about and why you did it that way.

The aim was to get right answers, so to be good at maths meant learning the process of how to do it. This was often a trick that was learnt by rote and had no reason, e.g. 'turn it upside-down and multiply it', or BODMAS (the order in which you had to do things: Brackets, Of, Division, Multiplication, Addition, Subtraction – if I've remembered it correctly!).

What some people say:

- I was no good at maths at school. It was all irrelevant.

- I could do maths at primary school, but at secondary school I didn't have a clue what they were on about.
- Maths has nothing to do with life. I don't use division, fractions, Pythagoras or quadratic equations in my daily life so I don't know why we had to learn them.
- We had these tables tests every Friday when I was seven and I was always bad at them. I never liked maths from then on.
- I was really good at maths until I did A level and that was a disaster. I failed it, so I lost my place at university.

No good at maths?

If you can:

- do the shopping;
- check if your bank statement is right;
- use a train timetable;
- follow a knitting, crochet or sewing pattern;
- cook a meal;
- play cards or Yahtzee;
- score at darts or snooker;
- do patchwork or cross stitch;
- decorate a room;

then you are using maths effectively – however much you may feel that you were not good at it at school.

Adults' memories of maths at school

Over the past six years I have been asking groups of adults about their perceptions of the maths they were taught at school. Over 93 per cent of those whom I asked said that they feared maths and didn't understand it. That is a terrifying statistic, but it is made worse in that most of the groups that I asked have been primary teachers, or student teachers. There are tales of humiliation such as being made to stand up and recite times tables but being so terrified that they could say nothing, resulting in the cane. There are tales of despair, of developing low self-esteem and of failure. These people tell tragic stories of having been punished – sometimes physically – for not being able to do their maths.

Schools are not that barbaric nowadays, but maths is still seen by some children as a subject that is to be feared. Many children under 11 who are enjoying maths at primary school will have changed their minds by the age of 13 and will go on to struggle with it in the hope of passing the exam at 16.

How maths is taught now

Some schools still teach maths by rote and by doing pages of examples from a book. This is seen as quite an efficient way to push children through exams, but it is in fact a very inefficient way of getting children to learn to think mathematically and to understand what they are doing in the longer term. This latter process takes a bit more time, and emphasises understanding more than relying on memory and teacher-taught methods, but in the end leads to children who are good at maths, who understand it, who enjoy it, and can use it in everyday life – as well as pass exams.

It is this slightly slower process, but with the much more satisfactory outcome, that is mostly seen in schools now, although the older the child, the more likely it is that she will still be asked to work from books and to learn something by heart.

The National Curriculum

The National Curriculum of England and Wales, and the Mathematics 5–14 used in Scotland, cover all the areas of maths that we might expect. These are:

- number (what we use to call arithmetic);
- algebra;

- shape and space (geometry);
- measuring;
- handling data (making graphs, statistics, etc.);

and also areas that were not usually much a part of maths before:

- problem solving (e.g. Can we plan a sports day for the rest of the school for the end of term? Can we think of ways to improve the playground?);
- learning to use and apply maths in everyday life;
- investigations (these are starting points for exploring mathematical ideas, and are explored in chapter 12).

There is more on the National Curriculum in the appendix on pages 192–198.

What this different way of teaching means for your child

Maths activities

What this means for your child is that there is as much emphasis put on activity and talking about maths as there is in writing maths down (particularly for the younger children). This activity, based around mathematical apparatus, is crucial for your child right up to GCSE level at 16.

This more investigative or 'open-ended' style of learning helps your child to think for herself and to see that maths is about patterns, relationships and communicating effectively, not just about getting the right answer. A few years ago a five- or six-year-old might have been asked to draw six houses and six balls and six cars, or to do a row of sums like this:

$1 + 5 =$
$2 + 4 =$
$3 + 3 =$

Those are OK as activities, but they require very little thought or creativity from the child and therefore the child does not engage very well with these tasks. If, on the other hand, you say to the child that

6 is the answer, so what is the question?

the level of involvement and thinking increases rapidly.

Here is an example of an open-ended activity that I have used very successfully with six-year-olds. Your child's teacher might give the children a number to explore and encourage the children to do this in any way that they want. These are some responses from my six-year-olds.

Making 6

$6 + 6 + 6 - 6 - 6 = 6$
$1 + 1 + 1 + 1 + 1 + 1 = 6$
6 lots of 6 take away 5 lots of 6 is 6
$96 - 90 = 6$
a billion and 6 take away a billion
20 + 20 + 20 + 20 take away 13 lots of 6, then add 4

You can develop this idea to think about 6 in many different ways in the classroom. I ask children to show 6 in any way that they want. Here are some things the children (aged six) in one of my classes did:

- got out an egg box and said 'this holds six eggs';
- said 'I'm six years old' and drew a birthday cake with six candles;
- put 6 grams on the scales;
- put the clock hands at 6 o'clock (another child put the hands at 6 minutes past 6);
- found (and marked with a coloured cube) 6, 12, 18, 24, . . . , up to 96 on a number line;
- put six hexagons together in a pattern;

- made 6 in as many different ways as they could with rods e.g. 1 and 5, 2 and 4, etc.;
- put out a cube and said 'this shape has got six faces';
- made 6 on the calculator (Sophie said '100 take away 94 is 6').

What is 'maths apparatus'?

Maths apparatus is an important feature of all classes – right through primary school – where the teacher has some understanding of the ways in which children learn maths. The children are not 'just playing'. They are doing something with their hands and that is a crucial part of their learning as they build, sort, count and all the other important aspects of using apparatus.

Each class will have apparatus something like this (the same apparatus is used right through primary school and in some secondary classes too):

- cubes;
- multibase cubes (Dienes apparatus for learning about place value, discussed on pages 89–92);
- an abacus;
- number lines;
- calculators;
- various geometric shapes (pentagons, triangles, hexagons, etc.);
- scales and weights;
- measuring tapes, metre rules and many other types of instruments for measuring such as a depth gauge (how deep is the pond?), a micrometer (how thick is this piece of paper?), a trundle wheel (for measuring large things such as the school hall);
- and many more things such as buttons, shells, pebbles and anything else that the child can learn from.

The maths apparatus is put where children can have easy access to it all the time.

As well as the apparatus that each class has, there will also probably be a central store of more specialised apparatus that is borrowed by classes when they need it, such as a thermometer, a rain gauge, a teaching clock with synchronised hands, and a set of three-dimensional shapes.

The importance of 'doing' in maths

It does untold damage for teachers in the junior school to tell their seven-year-olds, 'you're in the junior now so we are not having any of that babyish apparatus to play with'. Children learn by doing. They do not learn by being told about things – they need first-hand experience.

Sadly, some junior teachers do still tell their class that maths apparatus is babyish. If your child is in a junior school that minimises use of cubes, number lines, scales and anything else that might help your child to learn maths more securely, it is something to take up with the teacher and the head teacher.

> An old Chinese proverb:
>
> 'I hear and I forget,
> I see and I remember
> I do and I understand.'

All teachers want children to understand, and 'doing', and talking about that 'doing', is a vital part of learning.

The crucial role of language

One of the things that you will notice is very different from the way in which we did maths as children is that a maths lesson often involves a great deal of talking. You might even hear an infant teacher talk about the children being 'at the language stage'. This is a good sign! Children learn maths through talking and doing something with their hands so that they can see what is happening and can form mental images in their minds. These mental images will help them to think in more abstract ways as they get older.

Learning new concepts

Maths has a great many different bits to it and you might hear your child's teacher talk about 'concepts'. A concept is something like understanding what 5 means; or understanding the role of place value in number work (that a 4 in the units column mean 4 units and a 4 in the tens columns means 4 tens); or understanding what is meant by a curved face on a shape such as a sphere. The child understands these concepts by gradually developing a language (spoken and unspoken) with which he can explain them to himself.

It is through language that children (and adults) gradually grope their way towards understanding concepts. If someone wanted me to understand modern physics – about space being curved and why $e = mc^2$ – they would have to tell me slowly and use lots of diagrams and take a great deal of time to explain it to me. Language would be the key to it all as my brain tried to struggle with ideas that are difficult and new.

That is what it is like for children when they are trying to understand maths.

Maths with meaning

'$2 + 2 =$ ' is not meaningful to the young child. If you say to a child of about four years 'what does two and two make?' they say 'two what?'.

But if you ask the same child 'what does two lollipops and two more lollipops make?' they know it is four lollipops. This child can visualise two lollipops but doesn't know what '2' means on its own.

So, making maths have some meaning to the child is very important and this is done in maths lessons by talking and by activities based around maths apparatus. (Of course some lessons might be quiet or even silent if the teacher is trying to assess the children or if they are doing something on their own.)

Questions parents ask

Why aren't they doing any real maths at school?

For the very young child at nursery and in the reception class, playing is absolutely essential and without play your child would not learn. Later on, when a new concept is introduced, or a new piece of apparatus (such as a new construction set or an angle measurer), this is almost always preceded by playing. This is about getting familiar with the apparatus (what can it do?) and about giving the child some time to grasp what this new idea might be about. So, a nine-year-old might be given a fraction game to play with. It is not just an easy lesson for the teacher, it is part of the basis for the child's future mathematical understanding. That nine-year-old will not understand $\frac{1}{4} + \frac{1}{4} + \frac{1}{2} = 1$ until she has done something like cutting an apple into a half and two quarters or has played a fractions game like those in chapter 10.

Why are my children playing maths games every day at school?

That is good news! Games are often the way that children really grasp what is going on in maths. The games that you play at home, (snakes and ladders and ludo) teach your pre-school child how to count. Games played at school are there because they teach and reinforce vital maths concepts. Without them your child will not learn maths very securely. In this book I have suggested many maths games to make because this is one of the easiest and most enjoyable ways in which you can help your child with maths.

Girls aren't as good as boys at maths, are they?

I have put this in chapter 1 because it is so important. I am amazed how many people think that girls are not as good at maths as boys! Girls can be *every bit as good* as boys. It is a part of Western culture that they are thought not to be and we need to work hard to dispel that myth. If you have a daughter there will be many bits throughout the book that will touch on this point. It's hard work to do things that work against our culture, but for the sake of our child we must take seriously the issue of girls and maths.

Boys tend to dominate school life. Even though I knew that teachers give boys more attention (this is widely observed in research), I still found that I often didn't give the girls a fair amount of time when I looked closely at my teaching. Boys dominate the computer and the construction area. As a teacher I found that it takes a lot of tact to wean them off the idea that they have a right to do that!

' *How can I know how well my child is doing?* '

The only real answer to that is to talk to your child's teacher. Chat informally to the teacher if you just want a quick reassurance, or ask for an appointment if you would like to talk in more depth. A good teacher is always willing to talk to parents. But they are busy people, usually with far too much to do and far too many children in the class, so be patient!

With my own children I found that if they were happy in their class, got on with the teacher, had some friends and seemed to be coping with the demands of being awake and active from 8.30 to 4.00 pm, then they would be learning. If any of these went wrong for any reason, there was too much stress for much learning to take place. My other rule of thumb was that if the teacher was pleased with them and they were progressing, there was nothing to worry about.

' *What is maths for?* '

Maths is very closely connected with many of the life skills that we want our children to learn at school:

- learning to communicate;
- learning to think creatively and flexibly;
- learning ways of tackling problems;
- learning to use modern technology such as computers and calculators.

I don't use maths at all in my life so why do my children spend so much time learning all that stuff?

When I have talked to parent groups, we have thought about the maths that we have done that day. Lots of parents will say 'but I haven't done any maths today'.

But as we explore what our day has involved, the maths begins to show. Almost always at these parents' meetings there is a great deal of surprise at the maths that we do without realising it.

- Getting up in time to have a shower involves estimating time.
- Backing the car out of the drive without hitting the gate post involves estimating distance.
- Cleaning the house involves careful planning to do it in the most efficient way, and the need to work out what there is time to do before fetching the children from school.
- The shop keeper ordering goods for the shop makes predictions of what will sell.
- When driving a lorry there is a constant need to assess speed, distance and time.
- DIY involves working out how much wallpaper and paint and what order to do things in.
- Gardeners might need to use the fact that a lawn feed is 1 ounce to a square yard and the plants should be 12 inches apart.
- Shopping involves calculating: What is the best value? Can I afford that? Do I have time to get to that shop?
- Cooking involves time, temperature, weight, quantity, estimating.
- Sewing involves measuring sizes and estimating.

> *Why are my children allowed to do calculations in any way they want? Why can't the teacher teach them the proper way?*

A good example of the ways in which maths is changing at school nowadays is to look at how we were taught to do certain things and to compare that with today's way of teaching. Subtraction is a good example because there are lots of beliefs in the 'right' way of doing it!

This example might show you what we mean. Do this in your head: subtract 16 from 42.

How did you do it?

In a group of people all doing that, there would be a great variety of ways of calculating. Here are some of the ways one group of parents used:

1. Saying that if it was 16 from 46, the answer would be 30, but it is 4 less than that, so it is 26.
2. Using 'adding on' (rather like the way shop assistants used to count out change): from 16, add 4 to make 20, then add 20, which gets us to 40, then another 2 to 42, making 26.
3. If it was 16 from 40 it would be 24, so add the other 2, which makes 26.
4. 16 from 36 would be 20, then 37, 38, 39, 40, 41, 42, making 6 more.
5. Some parents wanted to write it down like this.

   42 –
   16

There were three different ways that this group did that written calculation! Some parents confessed that they were so flustered at having to do something in their head that they couldn't do it. A great many parents – and teachers too – develop panic about maths. We are trying not to pass that on to the children.

In all, we counted 11 different ways of doing that calculation. Children also have a great variety of ways of doing things in their head. These are their own methods that they have found for themselves. Having their own strategies for working things out makes children confident with their ability to handle numbers.

Encouraging children's own methods

What will be different in maths at school now will be that your children will tend to develop their own ways of doing calculations. We tended to be taught that there was one right way to do them but in fact there are many ways to do them. There isn't one right way. One of my children was persistently told to do her subtractions starting with the units. But she was much more confident with her own way of starting with the hundreds or thousands.

The changing needs of the world

When we think of the changes that have taken place since we were children, and when we see that the pace of that change is increasing, we can see that when our children are adults, the world will be very different. Their lives will be linked even more closely than ours are with the microchip, and computers will be a part of their everyday lives.

The flexible thinker

Because maths plays a most significant role in our everyday life, teachers are aware that the maths of their own education will not be sufficient for the citizen of the 21st century. Communicating in a variety of media, and having a flexible approach to tasks will be crucial survival skills for our children. There is nearly always more than one way to do something.

Becoming a mathematician

Many teachers think that the greatest thing that they can give your child as a future citizen of the 21st century is a good grounding in mental mathematics, a clear understanding of the meaning of mathematical ideas and a confidence in his own ways of working things out. That is why you will find that the traditional ways of working things out that we learnt are not taught much now, and computers, calculators and mental methods are used whenever possible.

Some dos and don'ts for parents

1 Don't say 'I was no good at maths so I don't suppose she will be either!'.
2 Don't say 'you're no good', 'you can't do it', 'you're hopeless'.
3 Don't make maths into a chore.
4 Do make learning at home fun and base it on play where you can. Children love to play – that is the way they learn.
5 Do build confidence and self-esteem by being positive and supportive.
6 Do praise your child for what she has done.

I have arranged the chapters of this book so that we start with maths for babies and then work through the age groups, ending up in the last chapter with your child going to secondary school. The exception is that right in the middle is chapter 7 and this has suggestions for maths activities that you can do as a family with children at different stages.

CHAPTER TWO

Helping your baby to learn

We give our babies a great number of non-verbal messages about their world. The ways in which we hold them, talk to them, make eye contact and behave towards them, all tell the child how much we value them.

Our home is the environment in which the child learns about the world. If we sing to them, read them stories and talk to them, we are immersing them in language and surrounding them with the basic tools of learning.

There is a very helpful section on 'How to talk to your baby' (page 15) in *Talking and Your Child*, a companion book in this series.

Parents and teachers often say to me that it is not maths that is important in the early years, it is language. Therefore they should concentrate on that, not on maths. There is a lot of truth in that – but I don't think that therefore young children shouldn't do maths! The maths for the young child is embedded in language.

The vital role of language in maths

Language is a basic tool for the child. It underpins all learning and it is as important to learning maths as it is to learning how to read.

It is language that is the means which young children come to understand the world around them. It involves not just words, but other sounds and non-verbal communication such as body language.

The earliest mathematical ideas are based on language:

- Learning to count 1, 2, 3, 4, 5 is based on talking and listening.
- When we have counted we say 'there are five fishes'.

Many ordinary words are basic to later mathematical development.

- The fish is *in* the water.
- The doll is *on top of* the cupboard.
- Let's go *down* the stairs.
- Let's build a *tall* tower with your bricks.
- You've got *more* apples than I have.

These words are learnt during talking, listening, singing and reading with your child and the importance of this early language development cannot be stressed too much.

Counting with your baby

Counting is only one aspect of mathematical learning but it is a very important one. Get into the habit of counting things out. As you put out the soup bowls, count them, 'one for you, one for me, one for Nicky, one for Sarah, that's 1, 2, 3, 4'.

I used to count out the vitamin drops – five every day. I remember people telling me I was crazy to do this, but I think it is never too early to start. Here are some examples of counting with a baby.

- One baby in a pushchair.
- Five bricks in the tower: 1, 2, 3, 4, 5.
- One cup of juice for you and two cups of tea for us.
- As you climb up the stairs, count them, one number word for every step.
- Count the clowns on the mobile. If you count them every night it will become a part of the daily routine and your baby will learn to recite the words.
- Use counting books at story time before bed. Make it fun and your baby will be learning.

Creating a learning environment

It is crucial that we create a positive and stimulating learning environment for our baby. As well as talking to them we need to play with them and involve ourselves in their play. Their surroundings need to be interesting and colourful and this need not cost much – it is easy to stimulate a baby with ordinary everyday objects.

Making a baby mobile

It is important to give a baby toys to look at and play with in the cot. From the very first weeks mobiles can be used to stimulate a child and to start to meet his needs to find out about the world around him.

Mobiles need not be expensive. They are very easy to make from materials found around the home. At first, just some silver paper or cooking foil, suspended well out of reach, will twist and turn, catching the baby's eye. I used to make fishes on a mobile from shiny gift tape with sequins for eyes. These were extremely complicated! Here are some much easier mobiles to make.

A big boat and a little boat

Cut the boats out of paper or thin card. Colour the boats with sweet wrappers, felt tips or crayons, and make the sails from aluminium foil. Put thread through the top of each boat and suspend from a drinking straw or wire. You will need to use lots of trial and error to balance the mobile – try tying the boats at different distances from the centre, or by making some boats heavier than others, if they are out of balance. Wire will survive being batted and kicked as the baby gets more and more determined to eat the mobile. Hardware stores sell wire in coils and I used to bend up the ends into a circle, as you can see in the picture, just in case a child got hold of the mobile.

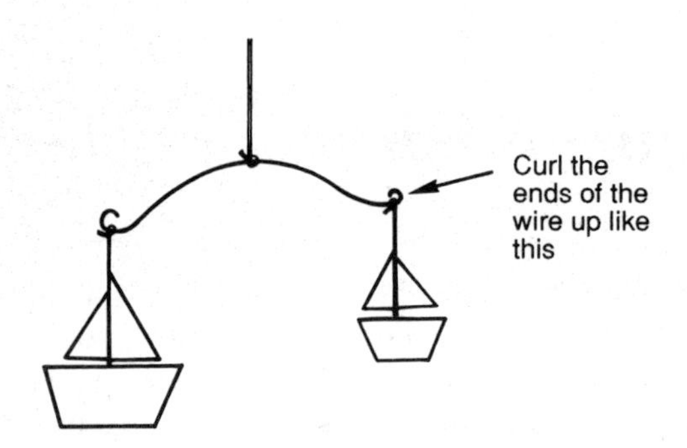

Some other ideas for mobiles

- One big green apple and two little red apples;
- Big Ted and Little Ted;
- The initials of the child's name;
- Teddies and cats, etc. can be cut out from bright wrapping paper or from birthday or Christmas cards.

As with any way of stimulating your baby, do be sensitive to her need for peace and quiet – she will soon tell you when she has had enough. Babies cannot take constant stimulation any more than we can.

The nursery number line

It's never too early to put up a 1 to 10 number line on the wall where the baby can see it from the cot. These are available in good toy shops. They need to be very bold in colour. Again, you can use brightly coloured wrapping paper or old greetings cards to make one. This number line makes a good topic of conversation at two in the morning when you are trying to convince her that she really does want to go to sleep. Count along the line.

1 teddy bear
2 chicks
3 baby birds, etc.

Keep it all soothing though, otherwise she will think you are ready for a game!

Mathematical toys for your baby

Anything bright and attractive to your child will encourage him to explore the world around him. A plastic colander and a wooden spoon suspended over the playpen will be kicked with great delight. Wooden spoons and bowls make great drums and plastic egg cups can be put under the bowl and made to disappear.

If you want to buy special toys, any kind of activity centre stimulates play and the toys suitable for a toddler can be used by a baby. (See chapter 3, page 35.) The baby will not stack her plastic beakers, but she is learning through her play that the blue one is the tiniest and the red one is the largest. You can hide her fluffy toy *under* the big one and say 'Where is your cat? Is it under the red beaker? Shall we look under the red beaker?'. Duplo Lego people will just be sucked at first, but you can play with them! If you make a point of playing for at least ten minutes with your baby and her toys during her most wakeful time, you will be teaching her valuable mathematical words – and that you value her enough to play with her.

Encouraging girls with 'active toys'

The toys that you give a child influence their play considerably. Lego, Mottik and other construction toys seem to develop spatial awareness that is so important for early understandings of geometry. (This is called shape and space at school.)

If a girl is just given soft toys or dolls, these are great for imaginative play, but rather restricting in terms of understanding about shapes that fit together and ways to build. Girls need things to bang and pull and build with, just as much as boys. So try to encourage relatives to give your daughter bricks, cars and construction toys. They may well tell you that girls don't like Lego. However, there are some people in the world who think that giving a one-year-old girl a plastic dustpan and brush gives her a worrying message! Of course there is a place for passive toys such as fluffy rabbits, and role-play toys like saucepans and dustpans, but it is when only the girl is given the rabbit and only the boy is given the more active tip-up truck that we are treading on dangerous ground.

Rhymes to say and sing

Very young children love rhymes. They love to hear the rhythms. They often rock, or, clap or kick their legs when they are being sung to. They enjoy and copy rhyming words, and giving children this feel for music and rhymes is very important for both maths learning and learning to read and write. It gives them a wide vocabulary and this has been shown to be crucial in any early learning.

Even the very youngest child will begin to pick up the language of number if rhymes and songs are a part of their everyday life. Some examples of number songs are given here and if these are used frequently, the baby will pick up mathematical words such as 'two', 'round and round', 'up and down', etc. as they pick up other baby words.

What we want is for the young child to feel happy and familiar with the language of maths from an early age. The time that most parents involve children in rhymes is at bath time, mealtimes, and that special time just before bed when stories are read and perhaps the child is sung to sleep. It is their general level of language ability that we improve and build on by talking, reading and singing to the young child. It will give them a good start not just in maths, but in reading, speaking, listening and writing.

Slowly, slowly, very slowly
Creeps the garden snail.
Slowly, slowly, very slowly
Up the wooden rail.

Quickly, quickly, very quickly
Runs the little mouse.
Quickly, quickly, very quickly
All about the house.
(Mime the actions with your hands.)

This rhyme is taken from *This Little Puffin*, compiled by E Matterson, Puffin Books. It is a wonderful collection of rhymes and songs which is widely used in nurseries and infant schools.

Round and round the garden,
Like a teddy bear, (your finger going around the palm of their hand)
One step, two steps, (walk your fingers up their arm)
Tickley under there. (tickle them under the arm)

One, two, three, four, five,
Once I caught a fish alive,
Six, seven, eight, nine, ten,
Then I let it go again.
Why did you let it go?
Because it bit my finger so.
Which finger did it bite?
This little finger on my right.

One, two three,
I love coffee,
And Billy loves tea.

Now you can see
One two three
I love coffee
And Billy loves tea.

*Two little dicky birds sitting on a wall, (*hold up one finger on each hand*)*
One named Peter, one named Paul,
*Fly away Peter, fly away Paul, (*put your hands one at a time behind your back*)*
*Come back Peter, come back Paul. (*bring hands out again*)*

*This little piggy went to market, (*hold their big toe, or thumb*)*
*This little piggy stayed at home, (*next toe or finger*)*
*This little piggy had roast beef, (*next one*)*
This little piggy had none,
And this little piggy went whee, whee, whee!
*All the way home. (*little toe or finger*)*

Engine, engine number nine
When she's polished she will shine.
Engine, engine number nine,
Ten will ride on the Toytown line.

(This can be said while clapping hands to the beat.)

Pat-a-cake, pat-a-cake baker's man
Bake me a cake as fast as you can.
Pat it and prick it and mark it with B
And put it in the oven for baby and me.
*(*Claps hands to the beat.*)*

Of course, you will want a book of traditional nursery rhymes to read. This will be one of your most used books, so choose a good one with bold illustrations.

Making up songs and rhymes

You can make up your own songs too. When my children were little we made up songs about the bath being filled, about bubbles, about the plastic ducks and Silas the seal and the many other creatures of bath time. It doesn't matter that your baby doesn't understand the words yet. She's picking up the idea that:

- the water is *in* the bath;
- she gets *in* the bath;
- put the fish *under* the water;
- there are *one, two, three ducks;*
- the blue boat is *on* the water.

Counting books for babies

There are a great many counting books for young children. If your baby develops a favourite from those you borrow from the library,

My own favourites include:

- *Teddy bears 1 to 10*, S Gretz, Collins.
- *How many bugs in a box?*, D Carter, Orchard.
- *There were 10 in a bed*, P Adams, Child's Play.

that is the one to buy him. If he is still into chewing books, some counting books are made in cloth or board.

Ignore people who tell you it is silly to count with a baby!

CHAPTER THREE

Mathematical activities with your toddler

By about a year to eighteen months your child will be developing a few words that are understood and starting to talk and 'read' books. Although children vary enormously, by about two-and-a-half they can usually say the words 'one, two, three' even if they aren't actually counting accurately. They can tell you who is Big Ted and who is Little Ted and they can understand that the bricks go in the brick box and the books go in the book case. Tidying up is fun and gives children valuable experience in sorting into sets.

This toddler stage, before going to playgroup or nursery, is a time of rapid growth and change in your child. She seems to learn new words rapidly and plays avidly, learning about size, shape, and colour. Even if she is not saying much, your talking to her is assimilated and will shape her future learning.

Again, language is the key.

Developing a positive relationship

Playing with your child at this stage and making the positive relationship that all children need in order to feel secure in the world is our first aim. If, as well as that, we can develop the feeling that numbers and other aspects of maths are fun, that is a bonus.

As with any stage in a child's development, don't force your child to learn. If you make it all into a game and play with your child, she will learn. If you insist that she learns, she will switch off and home learning will become a no-go area for your child. It is very sad to see that happening with children.

- Keep it fun.
- Stop when they have had enough.
- If in doubt, back off and do something different.

Helping your child to learn the language of maths

It can hardly be stressed enough that one of the most important things that you can do for your child is to talk to him. This is true for any time in his life, but the toddler stage is the time of rapid growth in vocabulary and understanding.

All the mathematical ideas that your child will need to be successful with maths at school are embedded in the language of play in the years before schooling starts. That is partly why nursery education is so valuable. But it is very important to realise that the close relationship with the parent and the language learnt at home is even more important and crucial to the child's development.

You will find more about this in *Your Child from 5–11*, another book in this series. Chapter 4, 'Communication within the family', is especially useful.

Young Children Learning: Talking and Thinking at Home and at School, by Barbara Tizzard and Martin Hughes, Fontana, is a fascinating account of a research project into conversations between four-year-old children and their parents that reveals the vital role of parents in the language development of pre-school children.

Here are some ideas for talking.

- Discuss the shopping list on the way to the shop. 'We need some cornflakes, milk and bread.' Encourage your child to spot them on the shelves. As your child gets older you can make this more demanding. 'What colour is the plastic top on the milk we buy?' 'Which size of cornflakes shall we buy?' 'Let's see how much this bread costs.'
- Involve your child in cooking, cleaning, journeys, gardening, planning where to go for an outing, choosing a plant to take to Granny (shall we take this one with the big pointed leaves, or this one with the little rounded leaves?), feeding the baby, tidying up, choosing a story to read, planning what to eat (how many bits of toast do we need?), setting the table (have we got enough plates?).
- Involve your child in selecting clothes for the next day. Will it be a hot day? Will the pink T-shirt go with the red trousers?

The more you talk to your child, listen to her and encourage her to take part in practical activities, the better she will understand mathematical ideas. You will be able to see how she is thinking as you listen and watch, and you will be privileged to see the exciting process of a small child learning.

What kind of talking?

You are the most important teacher your child will ever have and if you talk to him, he will learn in the natural and enjoyable way that all children down the years have learnt. If you want to think more closely about what kind of talking you can do with your child, here are some ideas.

- As you involve your child, *give him information* about what is going on. 'We must go to the greengrocers last so that we get all the other shopping before we have to carry the heavy potatoes home.' 'I'm going to soak this washing first because there are tomato stains on your dress from our spaghetti yesterday.' 'We need to ring Gran up because she went to the doctor today and we can ask her how she is.'

- Aim to make all the talking you do *enjoyable*.
- Keep building up your child's *self-esteem and confidence*. 'You did that really well.' 'Oh, that's very good.' It is from this constant praise and encouragement that he will learn to become a confident and independent thinker.

Important mathematical ideas at the toddler stage

Help your child to:

- sort things into groups: all the red things, all of your things, all the bricks;
- understand opposites: 'The soup is too hot to put in the fridge. We must wait till it has cooled down';
- appreciate pattern: 'Can we find the other sock with spots on?';
- start to understand about measuring: 'This shirt is much too small for you now. Look, your arms are too long for the sleeves'; 'We need to put in more milk to make the mixture thinner';
- matching one to one: 'Can you put a spoon beside every bowl?'

Important mathematical words

behind, in front, on top, beside, underneath, full, empty, half-full, more, less, bigger, smaller, fatter, thinner, long, short, huge, middle-sized and tiny.

If you can weave these words into your conversations with your child you will be building a firm foundation for her mathematical learning at school. Much of the maths at infant school is making sure that children understand these words. Without them they cannot grasp the maths they will learn at age seven and eight.

Learning to count

Maths is much more than learning to count, but counting is basic to children's future secure learning. It is important to note that children can say one, two, three, four, five, before they can count with a one-to-one relationship. What that means is that the child can recite the list of numbers (or what teachers call 'one-to-one correspondence'), much in the same way as he might recite any list of well-known words such as 'drink of juice', but he is not matching one number word to one object.

If you are counting things regularly with your child, and carefully matching one number word to one object (such as moving the bricks as they are counted, or taking slow and counted steps up and down the stairs), then your child will move from reciting the number words to actually counting.

Some children come to school at five unable to count accurately and your child's teacher will be ready for that.

Counting is a complex process

- Your child needs to be able to recite the numbers in the correct order. Even if you count the stairs from one to 15 every day from birth to five, she still might say them in the wrong order at age five. That's nothing to worry about. Getting this repeatable order right takes time and patience.
- She needs to learn that the last number in the count is the number we give to the set: 'One, two, three, four. So there are four bricks.' Children learn this with play and counting.
- She needs to learn that it doesn't matter how you arrange

the bricks, there will be four whichever order you count them in.

That's a lot of things for a two- or three-year-old to learn!

Activities to help your toddler to count

- Put out enough cups from your tea set for each bear.
- Give each teddy a biscuit.
- Set a place at the table for you, me and Elizabeth.
- You can have as many Smarties as fingers on your hand.
- Let's count the stairs as you go up to bed.
- Read counting books regularly and as a fun thing to do. If they enjoy the reading with you they will want to do it often and will learn without you having to pressurise in any way.
- Making your own counting book from pictures and birthday cards is fun.

- If you didn't have a number frieze when your child was a baby, one would be valuable now on his bedroom wall where he can lie in bed and look at it. If he did have one as a baby, you might want to put up another one or add more numbers to the existing one. Don't get rid of the old one as it may be a part of his night or morning activity to look at it. Count along the number frieze as a part of the bedtime ritual if you want to.
- Once you get into the routine of looking for opportunities to count, you will see them everywhere:

How many apples in the bag?
How many eggs in the box?
How many packets of cereal do we need to buy?
Let's count your teddies.
Let's count how many ducks are in the bath.
How many tea bags do we need?
How many bottles of milk are there in the fridge?

Activities around the house

- Sorting the washing. This is a good activity because children need to be able to classify things and to be able to say what goes in the set. This is a basic mathematical idea: a five- or six-year-old will need to learn to put all the triangles in one set and all the rectangles in another. The toddler can learn this same idea by putting all the clothes for the dark wash in one basket and all the light-coloured clothes in the other.
- Putting the wet clothes on the line is a time for counting. 'How many shirts?' 'Can you see the sock that matches this one?' 'What is the smallest thing on the line?'
- Sorting the dry clothes has a clear purpose. 'Can you find the missing sock?' 'Put all Clare's things in this basket and all Mum's things in this one.'
- Sorting anything out into colours is a good activity. If your child has learnt her colours before she goes to school that is a great help.

Games to play

You don't need to have any props to play a game. You can play a mathematical version of 'I spy' as you are doing other things in the house or out shopping. At first make the clues really easy:

'I can see something tall and green' – a tree.
'I can see lots of these and they are red and round' – apples.

You can develop this game into spotting the tallest thing you can see, or the fastest thing. Or you can say, 'there are three of these and they are red, amber and green' – the traffic lights.

'Simon says' is a good game to play when you are busy with something but your child wants to play. He has to follow your instructions when you say 'Simon says'. For example, if you say 'Simon says put your hands in the air' he should do it, but if you just say 'Put your arms in the air' and he does it, he is out.

This is a good way to teach position words: 'put your arms *behind* you', 'put your hands *under* the table'. You can also convey ideas of speed: 'move very slowly', 'jump up and down very fast'.

Playing with your child

Some useful mathematical play need only take a few minutes as you sit down with a cup of tea.

- See if you can build a tall tower with your bricks. Make it as tall as Sarah.
- Here are three dolls. Can you give each doll a ball?
- I've got three bricks. Can you put out a set of more bricks?
- I've got two red beads. Can you find two green beads?
- Let's count the biscuits . . . one, two, three.

Rhymes to say and sing

Continue with all the rhymes that you said when she was a baby. 'Two little dickey birds' begins to have a bit more meaning now and you could make a picture of the two birds with your child, or watch the birds out of the window and try to count them. It is good to try to count anything and of course there will be times when you want to say 'lots of birds' or 'too many to count'. That's an important mathematical idea.

Making pictures of the rhymes is a fun thing to do on a rainy day when a trip to the park is impossible. You could put the words and number symbols on the picture and stick it up on the bedroom wall. Then follow the words and numbers with your finger when you sing the rhyme together at night.

Hickelty Pickelty my black hen,
She lays eggs for gentlemen.
Sometimes nine and sometimes ten,
Hickelty Pickelty my black hen.

One, two three, four,
Mary at the cottage door,
Five, six, seven, eight,
Eating cherries off a plate.

One potato, two potato
Three potato four.
Five potato, six potato,
Seven potato more.

Five little squirrels sitting in a tree
The first one said 'what do I see?'
The second one said 'a man with a gun'
The third one said 'let's run, run, run'
The fourth one said 'let's hide in the shade'
The fifth one said 'I'm not afraid'
Then 'bang' went the gun and away they all ran!

Peter hammers with one hammer, one hammer, one hammer.
(Child sits and taps the ground with one fist.)
Peter hammers with one hammer this fine day.

Peter hammers with two hammers, two hammers, two hammers,
(Child taps the ground with two fists.)
Peter hammers with two hammers this fine day.

(Repeat with three hammers – two fists and one foot,
four hammers – both feet and both fists, then five includes the head as well.)

Peter's fast asleep now, sleep now, sleep now,
Peter's fast asleep now, this fine day.
(Child puts head on hands and sleeps.)

Peter's wide awake now, wake now, wake now,
Peter's wide awake now this fine day.
(Tap all five hammers again.)

Books for your toddler

Many of the stories that we read to our children have mathematical ideas in them.

Of course, when we read to our children it is the love of the story that we are mainly aiming for, but it is a good thing to bring out mathematical ideas as well.

Choose a few good counting books and read these when reading other stories. There is also a wide range of more mathematical books available about colour, shapes, sizes, etc. They are now available in supermarkets, but you get a wider choice from a children's book shop.

'The three little pigs' and 'The three bears' are obvious examples, but all the sequencing in something like *The enormous turnip* (M Shepherd, HarperCollins) is also mathematical (can they remember who came to pull up the turnip after the old woman?). Another real favourite in our house was *The elephant and the bad baby* (E Vipont and R Briggs, Puffin) which also has sequencing and is very memorable for reciting on long journeys.

A story like *The avocado baby* (John Burningham, Random House) has a number of ideas about size and strength, time and growth in it, and when you read stories like this, mathematical ideas become a part of your toddler's understanding of the world around him.

Toys for your toddler

Some well-known toys for toddlers have excellent mathematical value.

- Those stacking, different-sized beakers can be used for giving the teds a picnic. They can be used for towers, or for nesting inside each other. This one is the biggest, so this goes at the bottom. Which one shall we put on next?' They can be used in the bath for pouring, or used in the paddling pool outside. 'Which one holds the most?'
- Those wonderful Russian dolls that nest inside each other are a development of this idea.
- Bricks for building are an essential. My own children had some foam bricks first and then we moved onto beautiful wooden ones. At first it is enough to get one on top of the other – and then to knock them all down! Then bricks can be used for some of the most creative play you may ever see and will go on being used until they are seven and eight years old.

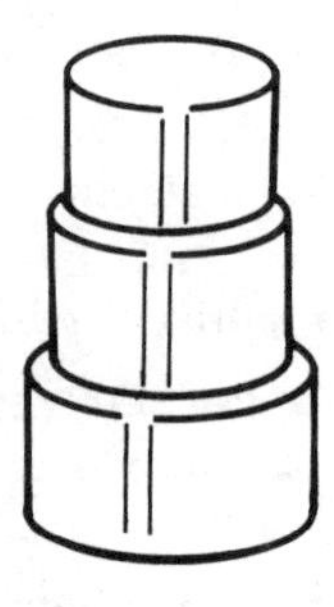

Imaginative play is tremendously important in helping a child in coming to terms with life and in practising and developing his vocabulary. You don't need to spend much money as big items such as a play house and the furniture can be made from supermarket boxes. Just giving your child a huge box can provide a source of play for hours. It might become a den, a spaceship, a zoo, a bed, a house or just be a hidey-hole to play peek-a-boo.

A big blanket or sheet over a table makes an excellent den.

Other kinds of mathematical play

Your child needs:

- water play in the bath – colanders, bowls and plastic spoons;
- construction toys – Stickle bricks, Duplo Lego or something similar;
- anything that can be sorted such as buttons, jewellery, or saucepans with different sizes of lids;

- some kinds of modelling material – play dough is ideal (there is a recipe for this in the appendix on page 189);

and if you can afford it:

- something on wheels such as a trolley or little tricycle (children cannot cope with pedals at first, so it can just be a push-along toy;
- a sand pit. This can be an old tyre filled with sand or a special plastic container. If you are into DIY you could dig a pit in the garden as this type is the easiest for the child to use and the easiest to keep covered to protect it from animals.

My rules for choosing toys

1. Is it safe? Market stalls and cheap offers may not give you the best value in the end. Look out for sharp edges, tiny pieces, breakable bits, eyes that pull out and the materials it is made with.
2. Is it fun to play with? We can't always know this, of course, but try to look through the eyes of your child. Beware of 'educational' toys that are boring!
3. Will it last? If your child passionately wants something specific like a doll's pram or a crane (this was my son's request for his third birthday), consider if there is a way of giving them what they would really like, whilst making the toy what I will call 'multipurpose'. Something with more than one use tends to have more lasting value – and can be very much cheaper in the long run.

I'll give a couple of examples of this from my own children's toys. When my son wanted the crane, I found that these were incredibly expensive if I was going to buy one that was strong and durable. Cheap alternatives looked flimsy. There is nothing so disappointing as a toy that breaks. So I bought a large pack of plastic Meccano that included a book of instructions with a crane in it. Then I sat up late on the night before his birthday and made an amazing crane! It was a work of art! It was on wheels, had a real string and hook that could lift other toys and was a great success. It lasted as a crane for some months and then it became adapted, unscrewed and made into cars, tractors, wheelbarrows, etc. Although the initial outlay was quite a lot, the Meccano was a firm favourite for years.

The second example is a Christmas present that I bought for both of my children from a mail order firm. It was a set of wooden cubes with one side missing, that could be stacked to make a house, castle, beds for teddies, a garage, or anything else that they wanted. Again, the outlay was quite a lot, but it gave hours of imaginative play for five or six years until my daughter requested a doll's house and then we glued it together to make a more permanent structure.

The toddler stage is wonderful for maths!

CHAPTER FOUR

The pre-school child – maths for three- to five-year-olds

What your child can do at this stage

Many three-year-olds can use small numbers well; for example, 'there are three bears and three bowls of porridge', 'I've got two biscuits and James has two biscuits'. They can often mentally add up two elephants and two more elephants, but they may not know what you mean by 'what is two plus two?', or 'two and two'. As we have seen earlier, these ideas are too abstract and the child might say 'two more what?' to show that they can't think about just '2' – it has to be two *something*.

At this stage children can recite the numbers on their number frieze and will say 'three butterflies, one, two, three'. They may not yet have grasped the concept of one-to-one correspondence (see page 29 in chapter 3), so a group of five raisins might be counted 'one, two, three, four, five, six, seven'.

They will know some position words, for example 'in', 'under', 'beside', but probably not 'left' and 'right'.

More mathematical words and ideas

- comparison words such as 'bigger than', 'more than', 'older than';
- measuring words such as 'tall', 'small', 'fat', 'thin';
- prepositions (or position words) such as 'under', 'through', 'around', 'on top', etc.

The importance of play

Play is a child's work. Children need to play in order to learn, and the quality of the learning environment that we put them in influences the type of play that they can do.

You can read more about this in 'Which pre-school group' in chapter 9 of another book in this series, *Preparing your child for school*.

For many children at this stage it is beneficial to attend a nursery or play group at least once a week, but we can do a great deal at home to improve and extend their play.

Construction play

Construction play at this stage is very important for the development of manipulative skills and for future mathematical, scientific and technological understanding. These three subjects are taught from the reception class at school onwards and you can help your child by providing her with a variety of construction toys.

Many of these toys can be quite expensive, but they are a priority (although your child will also get hours of fun from 'junk' such as plastic washing-up bottles and empty cereal boxes). Usually, the more you have of a construction toy, the more you can do with it, so it is a good idea to decide which ones you will set out to collect rather than buying lots of different kinds.

Stickle bricks (or Early Learning Brickles) and Duplo Lego can be used from about 18 months and ordinary wooden bricks are used from babyhood for many years, so these are good ones to start with.

Imaginative play

Your child at this stage is developing the ability to come to terms with the world around her and imaginative play is essential in that process. You don't have to teach children to play in this way. They do it anyway, but you can do an enormous amount to improve the quality of that play and so develop vocabulary and increase enjoyment.

The toddler played with empty boxes and this is true of the three- and four-year-old too. But you will see considerable development in this play as your child gets older and you will help the process with a few well-chosen toys. These can include:

- a telephone;
- a tea set (matching, sorting, counting);
- play dough;
- small figures and animals to add to the brick and construction play;
- a cash till (or improvise with a cutlery tray or other box with sections so that money can be sorted). You might want to have plastic money, but I think some real money is better;
- a calculator for adding up shopping lists;
- a note pad and paper for taking telephone messages. (Put the pad by the phone and watch your child copy you. It's amazing! He will 'write' and make lists just like you do.)

You might want to make some boxes into more things for creative play, such as a doll's bed, a cooker, a cupboard, a table for tea-parties. These items are expensive to buy in the shops and I think that the money is better spent on things that you cannot make at home, such as good construction toys.

The importance of games

Playing games is an important part of your child's mathematical education. Games teach:

- social skills, e.g. taking turns. These skills are crucial to success at school;
- counting and one-to-one correspondence: 'you have thrown a three so you can move three spaces – one (move one space), two (move another space), three (move another space)'. This movement of the counter helps the child to match one number word to one move along the board;
- the language of number ('more', 'less', 'fewer', etc.);
- matching (e.g. the number of dots on the dice is matched to the moves on the board).

Remember that it is the *game* that the child is interested in, more than the maths. Do remember that if you set out determined to teach some maths you might not succeed, but if you set out to have fun and play a game, learning will take place.

Games for three- to five-year-olds

- picture dominoes;
- picture lotto;
- number dominoes (from about four);
- ludo and other track games;
- darts (the plastic sort that stick onto fabric);
- Connect Four (or some other four-in-a-row game);
- jigsaws;
- picture snap;
- pairs or Pelmanism.

How to make a simple snap or pairs game (Pelmanism)

Cut up pairs of pictures from catalogues or wrapping paper and stick them onto pieces of card. (For snap with young children it is best to have three or four of each picture, otherwise it gets boring.) If you use card from the back of something like a cornflakes box, make sure you stick the picture so that it covers the side of the card with the printing so that when the cards are face down, you can only see the blank card.

You can make a more difficult version of snap by using coloured paper to make two red circles, two blue circles, two blue squares, two yellow triangles, etc. You can vary the game and sometimes snap for colour (that is easy for a young child) and other times snap for shape (so the red triangle and the blue triangle are 'snap').

To play snap, deal out the cards so each player has the same number and so that no one can see the face of the cards. Each player in turn plays a card from the top of their pile into the middle of the table. Two consecutive cards that match are 'snap' and the first person to call 'snap' wins all the cards in the middle pile.

To play pairs, make sure that there are only two of each picture and put all these cards out on the table with the picture face down. Each player takes it in turns to turn over two cards and look at the pictures. If they make a pair, they keep them. If they are not they turn them face down again, keeping them in the same position on the table. The person with the most pairs at the end of the game is the winner.

Pairs is great fun (by far the most popular card game with my own two children at this age), and it aids concentration and shape recognition. You can start with just five or six pairs and work up to 20 or so pairs by the time a child is four or five, and you can include pairs that are difficult to distinguish, for example a pair of happy faces and a pair of unhappy faces.

How to make a simple track game

You do not need to spend a lot of money on games. They are very easy to make. You need:

- glue and scissors;
- a large piece of paper, stuck onto card cut from the back of a cereal box;
- either some stickers, or pictures cut from a piece of wrapping paper or from a book bought at a jumble sale (it is best not to cut these out while your child is watching – it may give her ideas);
- some kind of counters (dry pasta shapes will do) and a dice to play the game. If you feel that a 1 to 6 dice is a bit difficult for your child, colour three or four dry butter beans red on one side. Use these for the throw and move the number of beans that are red-side-up. Or you can make a simple spinner with a piece of card and a short pencil. (Some people use a cocktail stick instead of a pencil, but these are quite dangerous for young children.)

You stick down the pictures or stickers to make a simple track around the paper, with a 'start' and a 'home'. Here are some suggestions based on the games I have made with wrapping paper.

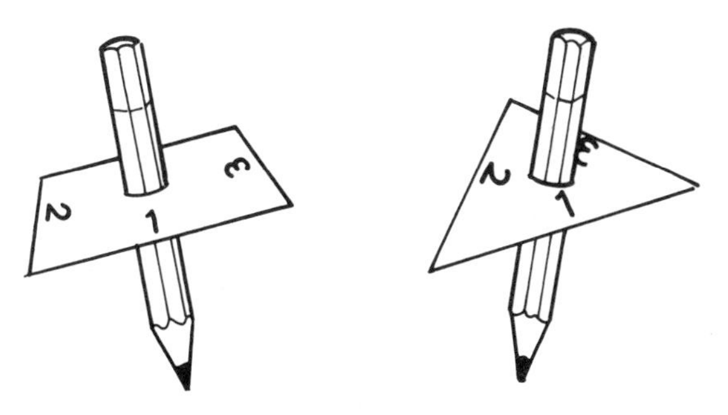

Using racing car pictures from wrapping paper (see over)
I have other ones like this made with different sorts of wrapping paper, featuring characters from television programmes and children's books. You might want to make the game a little longer if you have older children in the family. If you use a short game like this, you need to throw beans or use a triangular spinner because it would all be over too quickly with a 1 to 6 dice. (Use a dice though if it is 'just one more game before bed!')

Made with star stickers from a toy shop stuck down onto white address labels
You need ten or 12 labels in each row and you must have an equal number of stars in each row.

I used a piece of glittery wrapping paper for the background paper, stuck the labels and stars on, and covered it all with that clear sticky plastic you can buy in DIY shops. It has been one of the most popular games I ever made.

The star sticker game is different from the car racing games in that it can be used for a great many different games. You can play any rules you want and as your child approaches school age, he will get more and more interested in inventing his own rules. Here are some suggested rules.

- Play with just two beans. If you land on a star sticker, you win that many pieces of pasta. The winner is the person with the most pasta at the end.

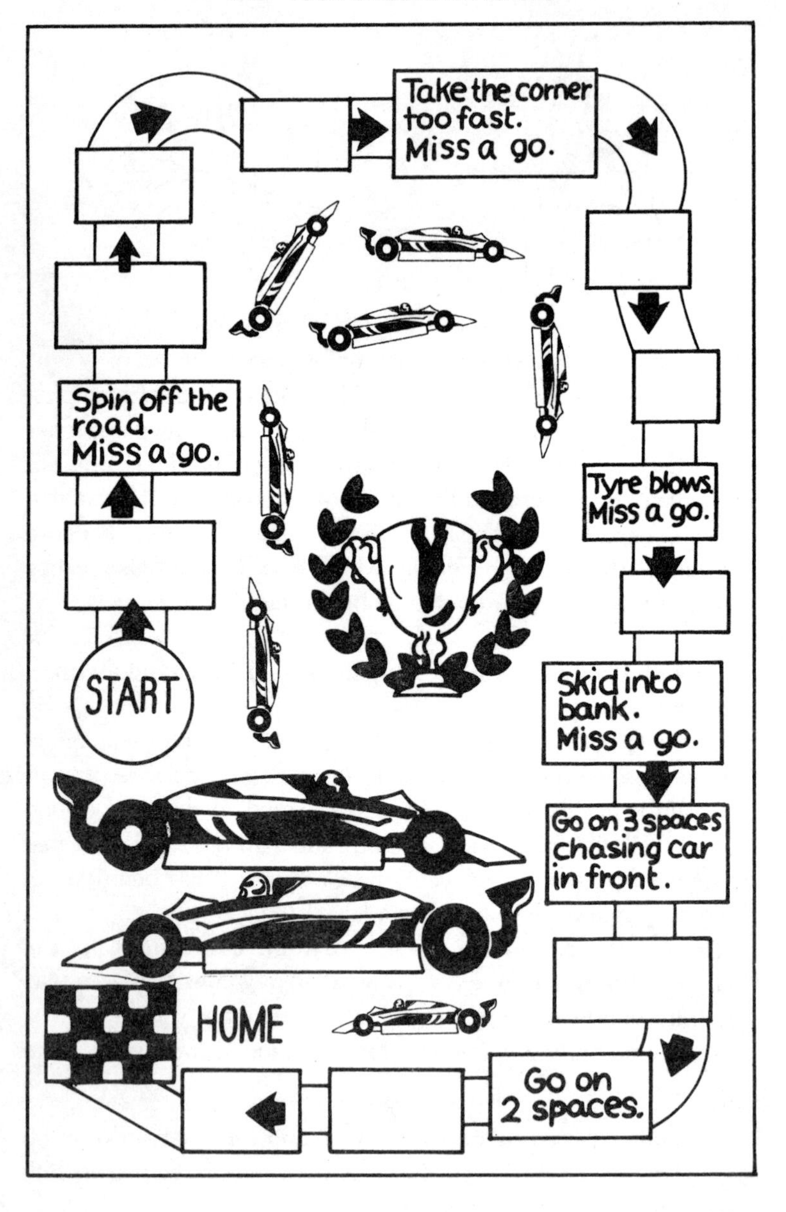
START
Spin off the road. Miss a go.
Take the corner too fast. Miss a go.
Tyre blows. Miss a go.
Skid into bank. Miss a go.
Go on 3 spaces chasing car in front.
Go on 2 spaces.
HOME

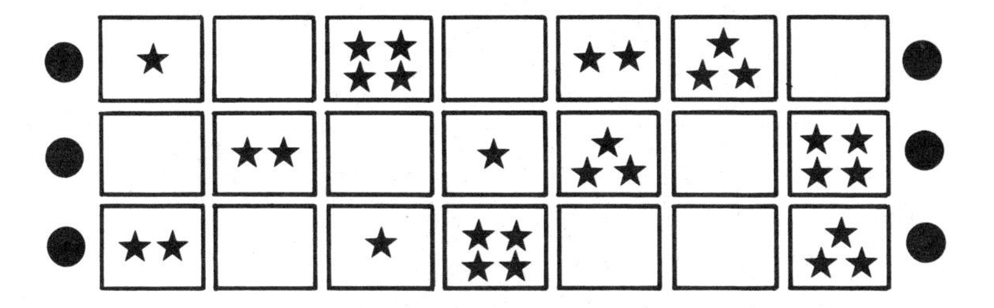

- Play with two beans, but this time, start with ten pieces of pasta. If you land on a star sticker you must put down that many pieces of pasta. The winner is the person with the most pasta left at the end. (This game is a play way to introduce subtraction.)
- Play with two beans, but if you land on a star sticker, you throw a 1 to 6 dice and win that many bits of pasta. When your child is approaching school age you can vary this to throwing the dice, adding that total to the stars and winning that much pasta.
- Play with no dice or beans, but with 'star cards'. Each player automatically moves a space at a time, but when they land on a star sticker, they take a star card and do what it says. It might say:

  'what colour is this?' (coloured paper on the star card)
  'which number is 1 more than 5?'
  'who is the tallest in our family?'
  'count to 10'.

  If the instructions are carried out correctly, the player moves on, if not she misses a go.

By playing this kind of track game your child will be developing his counting skills and his grasp of the one-to-one relationship (one red bean, one move along the track). If you make larger track games, or play snakes and ladders, he will be learning to recognise number symbols and getting a sense of the numbers up to 100. (It is not true that young children need to be kept to numbers under five. This incredible misconception is alive and well in many nursery schools and reception classes!)

How to make simple counting games

Target games

If you have some round counters you can make a simple tiddlywinks game. You only need a simple target that you can make from a piece of paper. Put the target on the carpet and flip counters on to it. If your child can't flip, then let her throw them. Keep the score, perhaps with buttons or bricks, or show your child how to use a calculator.

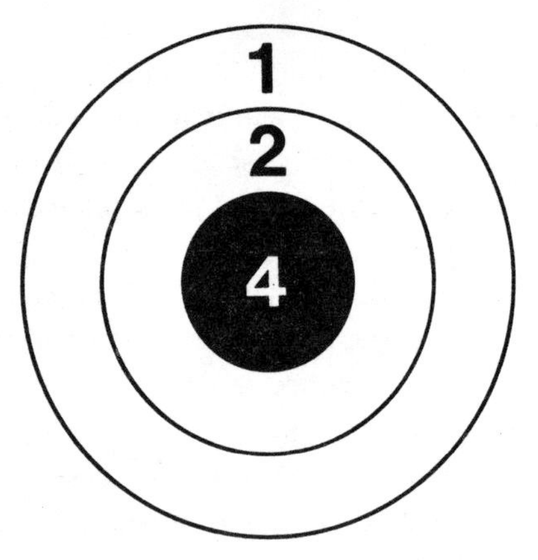

You can also play this outdoors with pebbles and a chalked target. (See also 'Maths on the beach' in chapter 7.)

Beetle games

These are easy to make from birthday cards. You need a picture stuck on to card for each player. Cut each picture into six pieces and label the bits. (The name comes from the game of 'beetle drive' that you might have joined in as a child.)

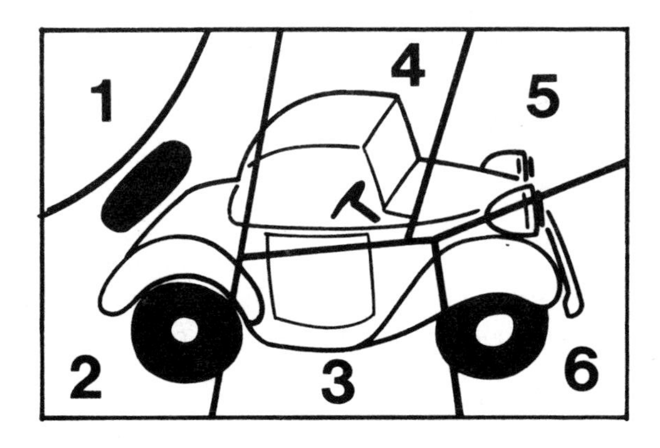

Throw a 1 to 6 dice and if you throw a four you can have the bit of your picture with a four on it. The aim is to complete your picture first.

Skittles

It is easy to make skittles from empty plastic bottles. At first the game can be to see who can knock down the most skittles. When children are older you can put a number on each bottle and score by adding up the numbers knocked down on each throw.

Games to teach number symbols

You can make a game of number snap with a set of cards with the numbers from zero to ten on them. You need three or four of each number otherwise it gets boring.

Vary the game a bit by making some cards with dots on, so three dots will snap with the numeral 3, or the word 'three'.

Number lotto is easy to make. Each player needs a base board and takes it in turns to pick up a card. If they have the number on that card on their board they can cover it. If not, they have to put the card back at the bottom of the pile.

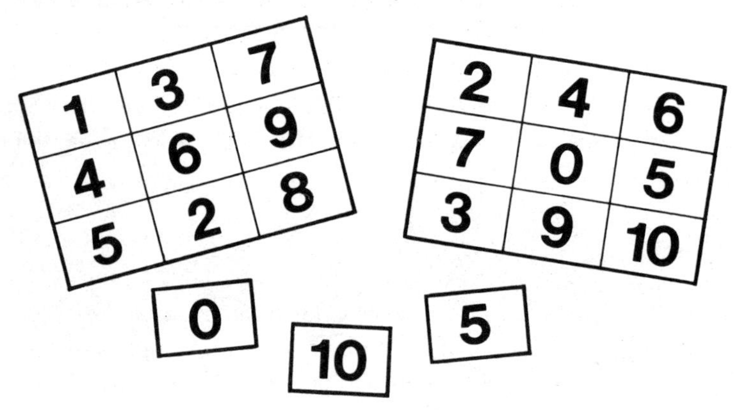

You can vary this like snap by having some cards with dots or words on.

Teaching your child to write numbers

When you show your child how to write the letters of the alphabet, show her the numbers as well. Remember:

- All letters and numbers start from the top.
- Keep it fun.
- Encourage your child even if he writes them wrongly; children grow out of reversing numbers by age seven or eight.
- Some children invent their own symbols for numbers. That is great – they are seeing that symbols are needed to communicate something. Encourage them. Don't say 'that's not a 2'.

This is the way that most schools use to teach numbers.

1 2 3 4 5 or 5

6 7 8 9 0

- Draw them for your child to trace over with a crayon.
- Make the numbers with play dough.
- Do dot-to-dot numbers for her to complete.

Conservation of number

You may hear your child's nursery teacher talk about this. What it means is that young children do not always see that a group of four sweets arranged in different way is still four sweets.

So a child counting this group will say 'one, two, three, four':

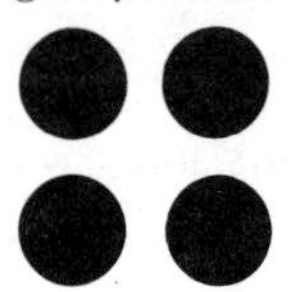

Then if you re-arrange the sweets to this grouping, the child needs to count again.

Children gradually learn that however things are arranged, they stay the same number and you can help that process by doing lots of counting with them. You can also make sets of dotty cards and get the child to put them in sets.

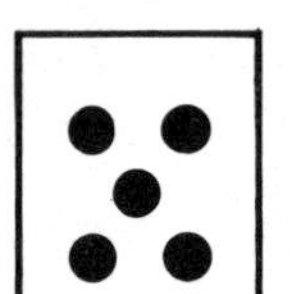

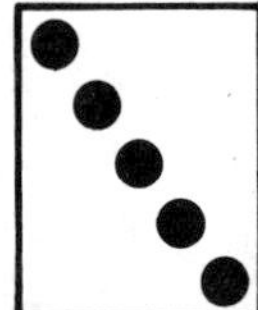

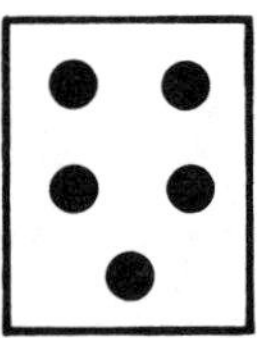

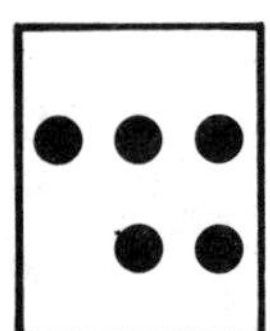

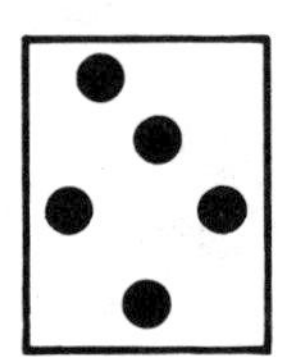

Making books with your three- to five-year-old

An important idea that you can teach your child is that books are fun and useful. This will help her in many important ways when she goes to school (the black squiggles are writing and it has meaning, books are for reading not scribbling in, and books are sources of information and they are fun). Many early mathematical ideas involve understanding certain words and it is these that you can teach your child through home-made books. Here are some ideas.

- Make a 'fat and thin' book (it might include a fat sausage, a fat cat, and a fat cow, a thin pencil, a thin sandwich and a thin house.) You might want to make the thin book a thin shape and the fat book squat and fat.

- Make a 'day and night' book (it could have the sun, going out to play and some shops for day, and stars, moon and a bed for night). Children need to understand the passage of time before they can tell the time.

A zig-zag book

- Make a book of numbers that are important to your child.

Children see a great many numbers around them – the bus number, road signs, car registration numbers, birthday cards, the weight of the contents on cereal boxes – and they enjoy exploring these numbers. Show your child how to put numbers into a calculator.

This is a good time to teach your child his address and telephone number.

- Keep a journal on outings and holidays. Record where your child went and what he did, and stick postcards and other treasures into the book e.g. a napkin from the motorway café, a bus ticket, the entry ticket for the zoo. This can be the beginning of map reading if you record where you went (you might draw a plan of the route to the park). It is important in appreciating distances (it is 100 miles to Granny's house), and it can give a child a sense of time if you record dates in the books. They make good bedtime reading, for example 'our trip to the zoo in the (season) of (year)'.

Your child may ask

'How old was I then?'

'How many months ago was that?'

and they can lead onto an understanding of seasons and the months of the year.

You can actually make a book about anything from the day the pipes burst to buying new shoes. Much of the learning from them may not seem very mathematical at first, but by involving your child in talking, you are helping her to develop a vocabulary that she needs for school maths.

Other things to do at home

Making a height chart

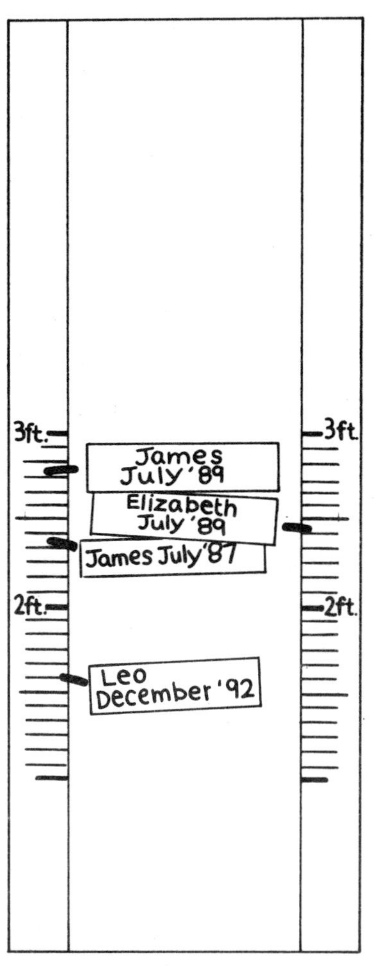

I still have a height chart from when my children were little and it is a real family treasure. These are very easy to make but remember that you want to write on it, so don't cover all of it with sticky-backed plastic. Covering the edges helps to preserve it.

If you can remember how long your baby was at birth, (usually about 20 inches) that makes a good starting point. Don't forget to put your child's name and the date by each measurement.

A child at 24 months is said to be half his adult height. I found that to be true for both of my children. So if your child is 3 foot tall at 24 months he might be on his way to being a 6 foot adult! Most clothes are measured in centimetres so you might want both imperial (feet and inches) and metric (centimetres) measures on your chart.

Learning to compare

Before children learn what measuring really means they need to learn to compare things (this is taller than this) and to put things in order of size.

The story of the three bears is great for this. You can dress up three teddies or other toys as Daddy Bear, Mummy Bear and Baby Bear. If you can find a bowl, spoon and chair for each bear, that adds to the enjoyment.

Other aspects of measuring

Measuring isn't just about measuring length, it is also to do with weight, temperature, area, volume, speed, angle and time. You can touch on most of these in ordinary play with your child:

- sand play (heavy, light, dry, wet, big sand castle, small sand castle);
- water play (half full, empty, running through the holes fast);
- cooking (the potatoes take 20 minutes to cook, the water is very hot);
- 'throw the ball into the bucket' (thinking about angle), 'flip the tiddlywinks or buttons on to the board';
- 'there is no more space on the table so you need to pack something away' (area);
- 'we don't have time for another game, it's 6 o'clock'.

More songs and rhymes

Ten fat sausages frying in a pan,
All of a sudden, one went bang.

Nine fat sausages . . .
and so on until you get to
No fat sausages, frying in a pan,
all of a sudden, the pan went bang!

Alternatively, you can get rid of the sausages two at a time, so you are counting back in twos:

Ten fat sausages, frying in a pan
One went pop and the other went bang.

Eight fat sausages frying in a pan, etc.

Five green speckled frogs,
Sat on a speckled log,
Eating some most delicious bugs,
Yum, yum,
One jumped into the pool,
Where it was nice and cool,
Now there are four green speckled frogs,
Glug, glug.
(Continue on until zero.)

Ten little monkeys, bouncing on the bed,
One fell off and bumped his head.
Send for the doctor, the doctor said,
'No more monkeys bouncing on the bed!'

Nine little monkeys, etc.

This old man, he played one,
He played knick-knack on my drum.

Chorus
With a knick-knack, paddy-wack, give the dog a bone,
This old man came rolling home.

This old man, he played two,
He played knick-knack on my shoe.
(Chorus at the end of each verse.)
This old man, he played three,
He played knick-knack on my knee.
This old man, he played four,
He played knick-knack on my door.
This old man, he played five,
He played knick-knack on my hive.
This old man, he played six,
He played knick-knack on my sticks.
This old man, he played seven,
He played knick-knack up to heaven.
This old man, he played eight,
He played knick-knack on my gate.
This old man, he played nine,
He played knick-knack on my spine.
This old man, he played ten,
He played knick-knack on my hen.

One man went to mow, went to mow a meadow,
One man and his dog, went to mow a meadow.
Two men went to mow, went to mow a meadow,
Two men, one man, and his dog, went to mow a meadow.
Three men went to mow, etc.
(Continue as far as you can stand it!)

These number rhymes are really crucial to your child's development and future number work. They really love them and they are great to make a journey pass more quickly or for some fun during odd moments in the day.

CHAPTER FIVE

Starting school – early years school maths

If you have memories of doing rows of sums in a squared-paper exercise book when you first went to school, you will find early years maths very different now. Your child will do sums but these will often be done mentally and in play situations. She can work out ten sums in the time it would take her to write down one. It is also more meaningful to her.

Maths at school – what to expect

The first maths that your child will do at school will probably be very similar to the maths at home, play group and nursery. A wise teacher does not put a child straight onto sums as these are pretty meaningless to a small child.

The sorting, matching, counting, building and mathematical rhymes of nursery will all continue but there will also be some progression towards recording mathematical experiences and talking in great detail about them. Here are a few examples of reception class maths.

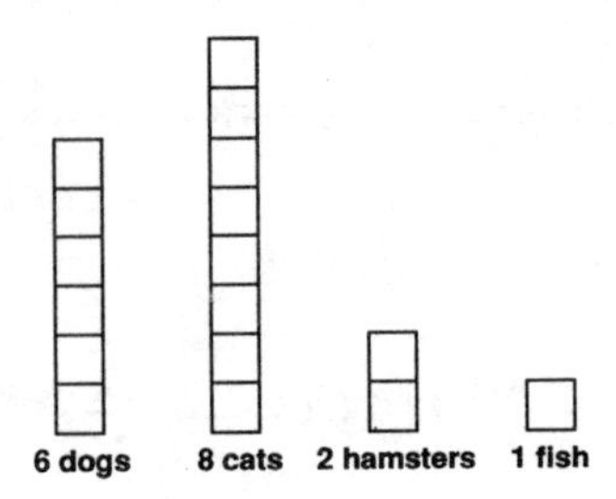

Making block graphs with towers of cubes 'In our class six people have dogs, eight people have cats, two have hamsters and one person has a fish.'

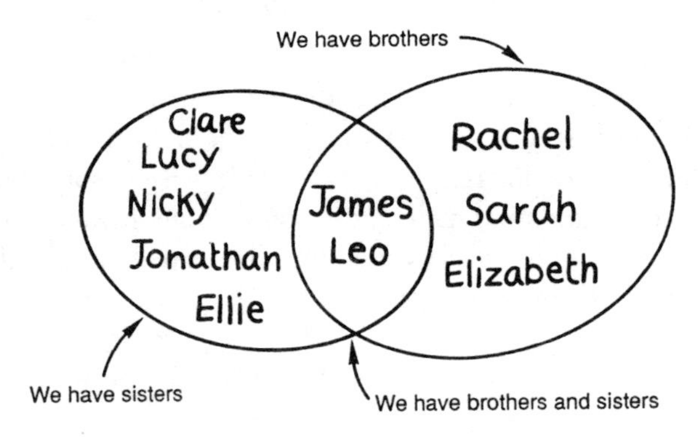

Making a Venn diagram 'Some people have sisters, some have brothers and some have brothers and sisters.'

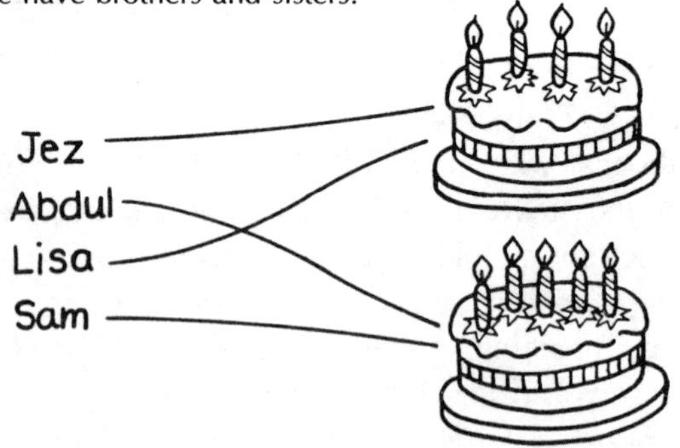

Matching (names are matched to the cake showing the right number of candles) 'We are either four or five.'

Mental maths

Teachers often put greater emphasis on mental maths (maths that isn't written down) than when we were at school. Mental maths for reception children might include:

'Who was the tallest when we measured everyone yesterday?'
'Which number is one bigger than 9?'
'What is 4 Smarties and 1 more Smartie?'
'Show me 5 fingers.'
'Who has built the tallest tower?'

and will go on to such things as:

'How many sides does a hexagon have?'
'Let's pick a shape out of the bag and see if we know its name.'
'What do you think is the probability of it raining today?'
'Which number is one smaller than 100?'
'Let's all count in 10s.'

The mental arithmetic that we did as children was different from this. That was about speed, it was written down, it was done on our own and it was about getting the right answer. I'm not saying that getting the right answer is not important – but our maths seemed to be *only* about getting right answers. The mental maths that we do now is much more about exploring different ways of doing things. It is a creative and exciting activity.

Maths is practical

Maths in the early years is very practical. If it isn't, from what we know about the ways in which children learn maths, your child will not learn securely and understand what she is doing. That is why there is less writing down than we did, and much more activity.

Children recording their maths

We get children to write down, or record, their maths for a number of reasons:

- to communicate something;
- to keep track of what they are doing;
- to learn to record systematically;
- to show parents what their child has done;
- to give evidence of their achievement.

What is crucial to realise though is that this recording is not an end in itself. It is the mathematical thinking that went on in the child's head before he recorded it that is so important.

Choosing a school for your child

When you are looking around schools to choose for your child, rooms of quietly working children all sitting in silence working at their books are not always a good sign! It depends what they are doing. Of course there is a time for silent working, but if that is the most usual way of doing maths, this may be a school that places a bit too much emphasis on what is written down. Because recording anything on paper is so slow in small children that the amount of maths they actually cover if they write all their maths down is very limited. Doing maths mentally and through talking and working with apparatus is very much quicker and the children usually learn much better.

There is more information on choosing a school in *Prepare your child for school*, another book in this series.

Preparing your child for school

Your child will cope with school if she is able to:

- concentrate on one activity at a time – listening to a story, or making play dough snakes;
- hold a pencil and, if possible, write her numbers;
- recognise colours;
- use mathematical words appropriately e.g. 'biggest', 'more than', etc. (see pages 28, 40 and 56);
- recite some number rhymes and, if possible, recite from 1 to 10;
- play with a calculator, and if possible, be able to recognise some of the number buttons;
- play games such as Ludo;
- count on her fingers.

Questions parents ask

❛ *My child seems to play a great many maths games* 

Maths games are so important that if your child doesn't play maths games at school this would be quite a serious omission from his education. You could compensate for this by:

- playing maths games at home;
- encouraging the school to buy or make some; or
- suggesting that the PTA has a fund-raising event to buy some.

❛ *My child is having trouble recognising numbers at school* ❜

- make sure she has a clear number line at home (see page 19);
- play games with numbers (e.g. snakes and ladders);

What maths games do

They:

- teach social skills such as co-operation and listening;
- provide a way of constantly repeating essential skills and concepts;
- show that maths is fun and something we can all do;
- help to develop your child's essential language of maths (many games your child will play at school require him to explain what he is doing, otherwise he may miss a go);
- provide a meaningful activity that your child can do while the teacher is busy with other children.

- look for numbers on buses, doors and on the cereal boxes at breakfast;
- play with magnetic fridge numbers: 'bring me the four', 'bring me the number that is your age';
- make some numbers that are her very own and play with them and trace over them each day, perhaps at the table at the end of the evening meal. Don't forget to include a zero, and form each number from the top down.

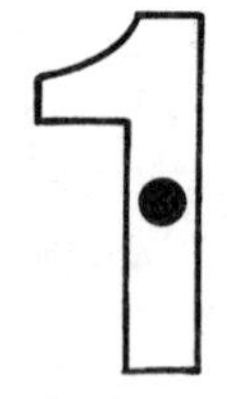

These numbers have the same number of dots on them as the number. So one activity can be to put four bits of dry pasta on the number 4

Why do children do so much matching and sorting at school?

Matching and sorting (putting things into sets, such as a set of things I like to eat, or a set of people with black shoes on) are essential parts of infant school maths because those activities underpin much of the rest of maths. If a five-year-old child can make a set of red things, he will be learning that a set is made up of things that are the same in some way. By six years of age, he will be able to make a set of pentagons and a set of shapes that roll down a slope.

Helping children to choose their own ways to sort objects can help children to see that they can sort in many different ways according to the criteria they select. So you can sort the contents of the larder into:

- tins or packets; or into
- sets of pasta and not pasta; into
- things I like to eat and things I don't like to eat; or
- sets with the same shape e.g. a set of cuboids, and a set of cylinders, etc.

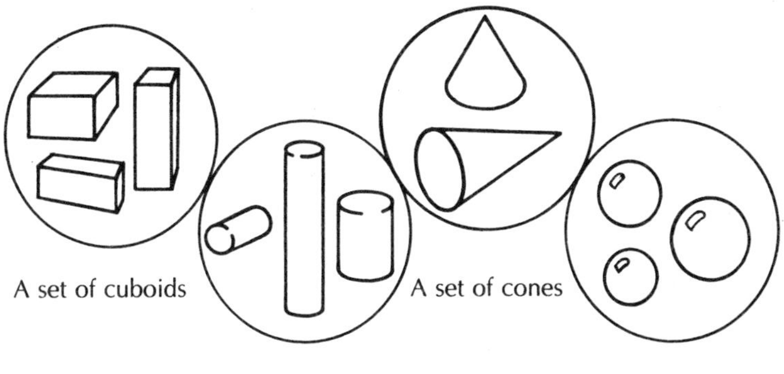

> *My child is starting to fear maths and he says he hates it*

It is rather alarming how easily children can start to lose confidence in their mathematical ability. It seems that one 'failure' or just a bit of teasing, and a child can learn to say 'I'm no good at maths'. We need to watch what kinds of messages we are giving our children. Are we inadvertently saying (maybe without using words but showing these in our attitude or in our actions):

- we love you when you succeed (interpreted by the child as *only* when she succeeds);
- we love you because you are on book 2 and the child next door is only on book 1 (possibly interpreted by the child as we will love you even more if you get to book 3 quickly);
- maths is so difficult that we don't expect you to succeed;
- maths is about getting the right answers and if you don't get them right that is bad.

If we get away from maths being about right answers, and look at the aspects of maths that are about exploring ideas or finding patterns, we can sometimes help a child to get back to enjoying maths.

What can I do?

- Keep being encouraging – 'you did that really well'.
- Keep building up self-esteem – 'wow! You are so good at maths!'
- Stay in touch with what is going on in the classroom. Is your child working with a group that is much faster at maths than he is? Would he work better in a different group? Did she miss something during an absence that is causing some loss of confidence?

- Play maths games at home, but don't force your child to play.
- Do some fun mathematical activities, e.g. have a Smartie party (see page 75) or have a teddy bears' picnic.
- Be led by your child. Children want to learn – but not always in the ways we expect them to.

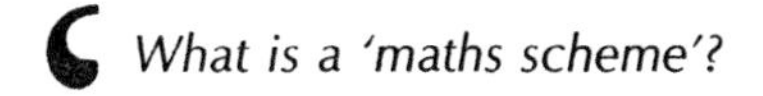

Many schools use a published maths scheme. This is a set of books for the child and a series of handbooks for the teachers that cover most areas of maths. The teacher's book will suggest activities to do with the children and the children at reception age usually have workbooks or cards to work from.

Some problems of a scheme

1. It is usually not very demanding of the children and consequently sometimes children do not learn much.
2. If each child is in a different place in the scheme, it is almost impossible for the teacher to keep tabs on who is learning what.
3. Children see maths as something that is about filling in empty boxes on a page and getting the right answer.
4. At reception age, schemes can be too much to do with colouring in and not enough to do with the maths.
5. Schemes do not usually allow for enough discussion about maths. Children *must* talk about their maths in order to understand it.
6. A scheme can put the emphasis on completing the page, but we want the emphasis to be on learning the concept.
7. When children are doing their scheme, they are not necessarily doing maths!
8. What a child writes down does not always reflect what went on in her head – and it is what goes on in her head that we are interested in.

Some advantages of a maths scheme

1. The children often enjoy a scheme.
2. It gives a structure to the teacher.
3. It is good for consolidation work – revising what has already been taught.
4. It can be useful as an activity to be done without the teacher when she is busy with another group.
5. It can include good maths games and well thought-out activities, and can form a 'backbone' of the maths in the class.
6. Teachers cannot be expected to keep coming up with new and different ideas so some kind of maths resources are needed.

'How can I help my child with money?'

Five-year-olds find that a 5p piece has the same value as five 1p coins very difficult to understand. I've tried all kinds of activities with small children to teach that concept (money dominoes, shopping games, etc.) but I was never really sure that these helped as much as handling real money and working out how to spend their pocket money.

If they have 10p to spend at the shop they probably will buy rubbish at first but they will get much better at it and will learn to save up for a bigger thing that they really want. A savings tin or a piggy bank is a great idea to help them to get used to adding up money. Even if they only sort the coins into sets that are the same they are learning something very valuable – to recognise the coins and to name them. It seems to me that children are noticeably less good at handling coins than children were a few years ago and I think maybe that is because so much shopping now is done through catalogues and payment is made by cheque or some kind of card.

I'm sure it could be argued that the child of the 21st century may not handle any money at all. That might happen, but equally there still might be parking metres, coin-operated telephones and fizzy drink machines so I don't think we have reached a stage when coins are obsolete.

My child is doing algebra – how is that possible when he is only five?

For most of us, algebra was that stuff we did at secondary school that was all about those *x*s and *y*s. It terrified the life out of me! I could never understand it. It seemed to be about endless rules that had absolutely no meaning or relevance to life. I was very worried therefore to find that I had to teach algebra to five-year-olds! If I couldn't understand it myself, how on earth could I ever teach it to a child?

What actually happened when I stopped panicking, was that I looked at what algebra for primary children involved and found that it was interesting – even exciting!

Algebra is about:

- finding unknown numbers – 'So if 15 is the answer, what is the question?';
- making repeated patterns – by threading beads (yellow bead, red bead, yellow bead, red bead) or printing (square, square, triangle, square, square, triangle);
- finding any kind of pattern in maths (in chapter 12 there is a 'growing pattern' that I have called 'cans of beans');
- clapping rhythms (this is about repeated patterns) and singing any song with a chorus. They have an element of algebra in them because the pattern is verse, chorus, verse, chorus.

So letting your child make threaded bracelets and necklaces and decorating paper crowns with stickers is algebra.

What can I do to help at my child's school? 

Most schools will be very glad of any help you can offer them. You could:

- play maths games with a group;
- cook with a small group;
- help to run a maths games library;
- help to run a toy library.

My child is at the beginning stages of learning her tables. How can I help her?

Multiplication is usually started at about age six or seven with putting things into groups or sets and with counting twos and tens. Here are some activities you can do with your child.

Shopping

- Eggs come in sets of six so how many boxes shall we buy? (Start with six, count on another six on your fingers, which makes 12. Will that be enough for the week?)
- We need 20 bread rolls. They are in packs of 8. How many will we need to buy? (Work it out with buttons or actually count them in the shop.)

Playing

- Put the mince pies in the oven trays, six to a tray. We have three trays of six. How else can we say that?

  3 sets of 6;
  3 lots of 6;
  3 groups of 6.

  (It is very important that your child learns all of these different ways of expressing multiplication.)

On journeys

- Let's all count twos. Two, four, six, eight, ten. When your child is secure with twos, try counting in fives or tens. Consolidate these for several months and don't attempt others if your child struggles with these early number patterns.
- There are some cows. How many legs do three cows have? Four, eight, twelve. (Use fingers to help with the counting.)

Pictures on the wall

One very effective way to help your child to see what multiplication is all about is to make some pictures for his bedroom wall. My son, age three, said to me one morning, '5 lots of 7 makes 35'. Completely astonished by this I asked him how he knew that. He said he liked to lie in bed and count his butterflies on his wall. These were arranged in five rows of seven.

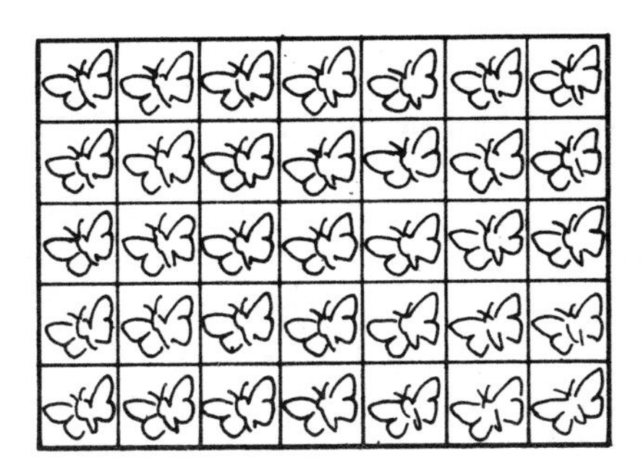

So now in school I use pictures arranged in rows and these are easy to make. Here are some ideas:

- Cut up some pictures of things that your child is particularly interested in (cars, boats, dinosaurs, horses, dogs, stamps, etc.) and arrange these on a large sheet of paper in rows. Five rows of five is a good one to choose because children like to count in fives.
- A calendar has days arranged in sevens and you could do a 'countdown to my birthday', crossing off the days each night or morning.
- Potato prints or rubber stamps in regular rows make a good picture (perhaps four rows of ten).
- Some wrapping paper is printed in regular rows and makes a quick and colourful poster.

Making links with school

It is very important that you make links with the school when your child is young. This helps to prevent problems from building up and you and the teacher can get to know each other and can discuss the child's progress informally. You obviously want to know if your child is happy and coping with school and the teacher will want to make a relationship with you as it helps her to understand the child.

Home/school maths projects

There are a number of home/school maths projects that schools take part in. The idea of these projects is to develop maths in much the same way as reading has developed over the years. It is usual for children to bring home a book to read so people began to say 'well, why don't children take some maths home?'.

We know the advantages of home/school reading. Children become confident readers much more quickly and it turns out that it is much the same with maths.

If your school does run a home/school maths project, your child will bring home sheets with activities like those suggested in this book. They are fun maths things to do together. For example, one sheet might have a simple activity about area. It might suggest that you lay sheets of newspapers down on the floor of a room to see how many sheets (not overlapping) are needed to cover all the floor. The school might ask for one sheet of the newspaper to be taken back to school with the results. This might lead to discussion at school that some people use different sizes of newspaper, and it will be accompanied by some other related activities on area at school.

This type of home/school co-operation can help your child to learn and enjoy maths, and you get to help in the process. It is important not to see these home/school projects as traditional 'homework'. They are so successful because when the parent and child work at something together, the child seems to learn very well – and parents learn how their child is thinking.

Other number activities to do at home

We can replace the 1 to 10 number line of the nursery years with a line that goes all round the room, perhaps colouring every tenth square in red, or colouring alternate strips of ten in red, then green (so one to ten would be red and 11 to 20 would be green and so on).

1 2 3 4 5 6 7 8 9 10	11 12 13 14 15 16 17 18 19 20	21 22

To help multiplication, as well as a number line, you could have a 100 square poster. You could make this with your child.

Divide a large piece of card or paper into ten rows by ten. (You could join together two bits of a roll of wallpaper.) Show your child how you are doing this. She may not understand how you are using the ruler or measuring tape, or see exactly how you are folding it, but she will learn through you talking and showing her what you are doing. Then put in the numbers 1 to 100 with a large felt tip, or with a child's paint brush. Make the numbers large and clear. Put this up on the child's bedroom wall where she can see it.
You can mark on the numbers in the two-times table in one colour, the five-times in another, and the ten-times in another. Only do these three tables at first.

1	2	3	4	5	6	7	8	9	10
11	12	13	14	15	16	17	18	19	20
21	22	23	24	25	26	27	28	29	30
31	32	33	34	35	36	37	38	39	40
41	42	43	44	45	46	47	48	49	50
51	52	53	54	55	56	57	58	59	60
61	62	63	64	65	66	67	68	69	70
71	72	73	74	75	76	77	78	79	80
81	82	83	84	85	86	87	88	89	90
91	92	93	94	95	96	97	98	99	100

Ideas for a party

A teddy bears' picnic

You don't have to wait until summer to have picnics. We had indoor picnics in our family and they were a great favourite. Just a few cuddly toys, a tea set, a plastic cloth on the floor and a few bits of food and children will be learning maths as they share, pour and count. Give them biscuits and squash to share out and read them a few stories such as *The Tiger who came to Tea* (J Kerr, HarperCollins).

You could develop this by inviting a couple of their friends around and asking them to bring their teddies. Give them perhaps some slices of apple, some carrot sticks, a pile of raisins and a small cake to share out evenly (give them a blunt knife or a pie-slice). They can put out portions for the teddies too, then, when the food is eaten, use the teddies for some mathematical activities.

- Which bear is the oldest? (Do they remember when they were given the bear? You might need to ask parents, but it is a good activity to give children a sense of time.)
- Which bear is the tallest? You could put them all in a line to measure them one against the other. If your child suggests getting a measuring tape, that is wonderful. Let them try to use it and if they ask, show them how to use it.

If you want you could make some little certificates or prizes for the winner of each category, making sure everyone wins at least one.

- Other categories you might choose are the shortest/furriest/fattest/baldest/prettiest/one with the smartest clothes, etc.

All of those categories include mathematical ideas and they will give the message that maths is great fun and we can all do it.

A Smartie party

You can get an enormous amount of maths from those little boxes of Smarties that you can buy in supermarkets.

Note

Some children cannot eat Smarties because of the food colourings in them. Look for healthier substitutes in chemists' shops, or have some chocolate buttons for these children.

Here are some ideas for five-year-olds. Everyone washes their hands and no-one eats a Smartie until all the games are over! That's tough so you might want to provide crisps and savoury sandwiches to keep them going.

- How many Smarties have you got in your box?
- How many of them are blue, green, red, etc.?
- Are the colours the same in each box?
- Are there always the same number in each box?
- How many Smarties have we got here altogether? (Use a calculator.)

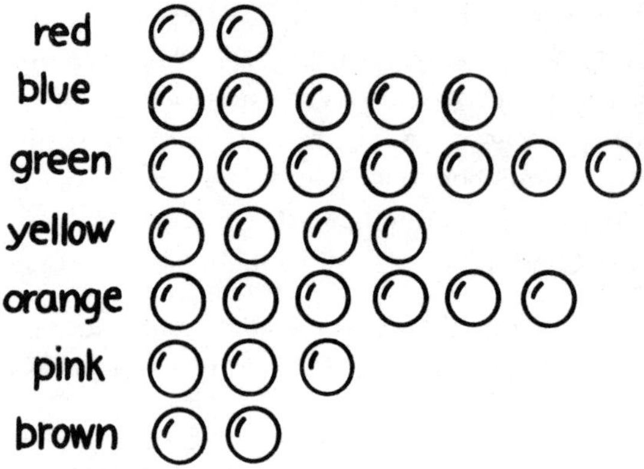

red
blue
green
yellow
orange
pink
brown

Sally's box had more greens than any other colours

With your seven-year-old you could be a bit more adventurous and consider whether it is better value to buy the little boxes or tubes and even work out with a calculator how much one Smartie costs! (I would advise you to do other costings with your child first such as how much it cost to buy all the ingredients for the pizza. Costing can get you into fractions of a penny and you need to be sure that you are not going to do something too difficult for them – not to mention that it is a bit mind-blowing for us to work it out!)

Other Smartie activities

- Make some paper hats with stickers that look like Smarties. Encourage children to make repeated patterns.
- Hide Smarties around the house and have a treasure hunt.
- Make up a Smartie song to the tune of 'Ten green bottles'. 'Ten coloured Smarties sitting in a box', etc. (Children are wonderfully inventive and will enjoy silly rhymes.)
- Play a guessing game. Blindfold each child in turn. They have to try to predict which colour of Smartie they will pick out of the tube. (You can surprise them by making up a tube that only has red and green ones in it.)

CHAPTER SIX

More things to do with your five- to seven-year-old

What your seven-year-old can do

Children vary enormously, but usually by age seven a child can:

- count reliably;
- talk about the passage of time (night time, winter, etc.);
- use a calculator reasonably confidently;
- work independently with some computer programs;
- discuss ideas about huge numbers and infinity;
- make a simple pattern (such as red, blue, blue, red, blue, blue, using cubes);
- recognise and name 2-D and 3-D shapes;
- use small numbers with considerable confidence and remember some number bonds (that means remembering useful number combinations such as $5 + 5 = 10$, $6 + 4 = 10$, $7 + 3 = 10$);
- add and subtract mentally with small numbers and use apparatus or a pencil and paper to do the same with larger numbers;
- understand simple ideas about measuring (I'm taller than Jubeen, but shorter than Salik).

Activities in the home

Teaching children about time

Children of seven are not usually good at telling the time but you can teach them about the passage of time, the days of the week, the months of the year and the seasons.

Teach your child

Thirty days hath September,
April, June and November,
All the rest have thirty one,
Except February alone
Which has 28 days clear,
And 29 on each leap year.

If you hang a calendar on the kitchen wall and use it to organise family events, children gradually absorb ideas about time. They can see that it is the dentist on Wednesday, swimming on Thursday and the library books are due back on Saturday.

Learning about seasons

It is easy to make a seasons calendar with your child and she will enjoy turning the pointer. You can use this to discuss ideas about probability: 'will it be colder outside in a month's time?' 'Do you think it will snow on Christmas day?'

The basic seasons calendar is a circle divided into quarters. Your child can do appropriate drawings for the seasons and you can add the months of the year when you think your child is ready to learn them.

My book of dates

This is great fun to make and is best done in a book with a fairly stiff cover as it could become a family treasure. You enter any dates that you can think of and perhaps add some family photos or mementos. Here are some examples.

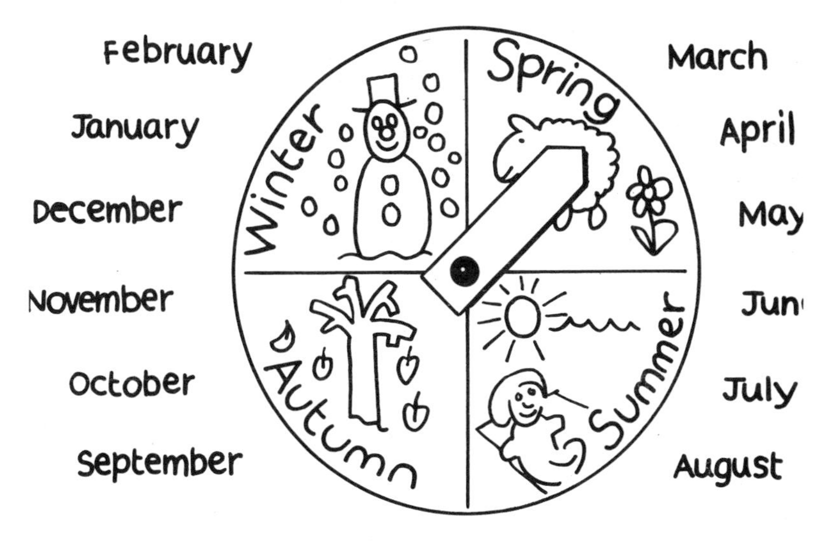

Grandpa was born in 1935.
Dad is 26 years older than me.
My house was built in 1972.
We went to Brighton for our holiday in . . .
This is me when I was . . . years old.
This is me on my first birthday in . . .

You might want to add photos of grandparents or even great-grandparents to the book, or draw a family tree.

A book just about birthdays is fun

You can add a birthday photo, some or all of the birthday cards, the child's height on that day, who came to tea, etc.

You might want to add the 'When I was one . . .' poem from A A Milne's book *Now we are six* (Mammouth) and any other birthday poems you can find.

Once your child is about seven you can do some birthday maths.

My book of birthdays

What year will I be 18?

How many more years is that?

What is the difference between your age and Nicky's age? ('Difference' is quite a difficult aspect of subtraction and I think it is best taught from ages. Most children know that they are two years older than a brother, so the difference between 7 and 5 is 2.)

Making a time line

Some kind of very simple line will help your seven-year-old to get some more ideas of time and the sequence of historical events. The line needs to be very simple at first, such as a piece of wool across the bedroom with events in history that your child knows about tied onto it in order.

The dinosaurs would be at one end and space ships at the other.

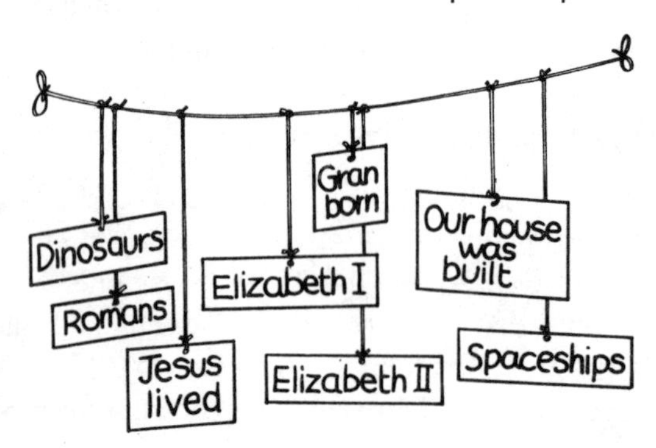

Elizabeth the First comes before Elizabeth the Second and the old house in the High Street was built in the time of Henry the Eighth. This will lead into ideas about big numbers ('that was a thousand years ago').

Helping children to add up

Games

If your child is struggling a bit with adding up, it is worth investing in some games that require adding, such as a dartboard that works with velcro (available from Early Learning Centres, see the resources list on page 191) or making some kind of target game (see chapter 7). It is easy to make a hoopla board with a bit of wood and cup hooks (the self-adhesive kind are the easiest). For hoops we used those rubber rings for preserving jars that you can get in a hardware shop.

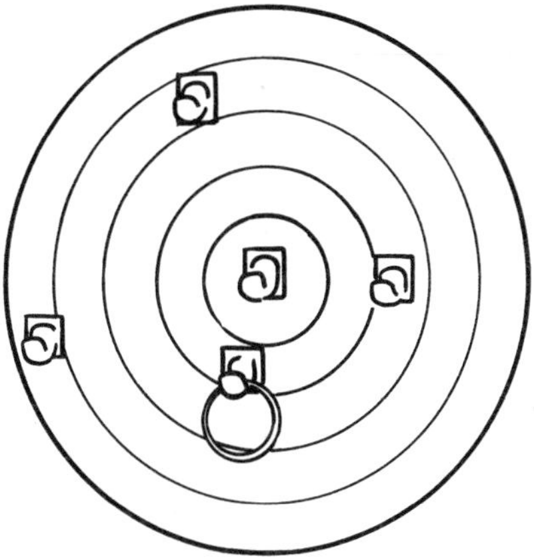

Other adding games that have scores are things like cricket and indoor games such as dominoes, Monopoly, skittles (see page 49) and magnetic fishy games that are expensive to buy but really easy to make. You need:

- a stick;
- some string;

- paper clips;
- a magnet (you can get these at a good toy shop and a horse-shoe one is usually the cheapest and easiest to use);
- a box;
- some paper and a pen.

Make paper fishes with numbers on (or with sums if you want). Make a fishing rod by tying the magnet to one end of a piece of string, and the stick to the other.

Give each fish a paper clip nose (to attract the magnet), put them in the box and take it in turns to fish them out. Your child can add up the totals of the numbers to see who has the largest score at the end. You can adapt the numbers for the age of your child. Try enormous numbers and use a calculator to add them up! If you number one fish '1 000 000' (a million) you will get lots of discussion about big numbers and infinity.

Adding on journeys

Other adding practice can happen in the car, or looking at traffic. Choose a number e.g. nine, and try to spot number plates that have digits that add up to nine (e.g. 252, 342 and 801).

If you are going on a long journey, decide on various things to see (e.g. number plates that add up to eight, a number plate with all the numbers the same, a 30 mile an hour sign, 100 written on a sign post, something with four legs, and a pub sign with a number in it such as 'The Three Fishes'.) The person who sees each thing first can cross it off her list.

See chapter 7 for more ideas.

Involve your child in the maths you are doing

A seven-year-old would get a great deal from adding the shopping up on a calculator, or from being involved in any of your financial working out. Can we afford to buy more daffodil bulbs, go on holiday, buy you a new pair of shoes? (Take care not to worry a child with adult financial problems.)

By seven a child needs to be handling his own money. They get very good at working out how much more they need to save up for something.

Remember:

- Your child will work things out in her own way. That's very important and you will be helping your child if you encourage it.
- Estimating is terribly important. Ask about how much they expect an answer to be. This helps children to know if they have pressed the wrong button on the calculator and to check their mental working out. If they get an answer nowhere near what they expected, something has gone wrong somewhere.

Cooking

Do let your child cook with you. There is so much maths in cooking, from buying the ingredients to working out how many potatoes you need for three people. If you can afford to buy them, balance scales are ideal for weighing, but they will get access to these at school. If you only have scales with a dial that gives the weight (it's all I had for my children) teach them to read the dial. What I did do was give away my old pounds and ounces scales and buy some that read in both metric and imperial measures.

You will be aiming at letting your child cook on his own, but he needs lots of practice with you first. Teach him about the oven heat, how to use the egg timer and any other timer you might have, how to measure out liquid in the jug, how to use the mixer, and simple basic skills such as breaking an egg and whisking it, rubbing fat into flour and cutting out biscuits and placing them on a greased tray.

I am always rather alarmed at the way that children seem to learn to cook cakes and biscuits first. That gives an odd message about the food that we encourage them to eat. It's just as easy to learn to cook scrambled egg, savoury popcorn, beans on toast and to make a salad, as it is to make fairy cakes. Having said that, of course, children love to make and eat biscuits and cakes! I just mean that we need not restrict cooking to sweet things.

Here are some recipe suggestions.

Cheesy hedgehogs

Cut up some cheese into cubes and spear these with cocktail sticks (take care that these are not swallowed). Cut an orange in half and stick the cocktail sticks into each half so that you have two hedgehogs. You can let your child invent a way to give each hedgehog eyes and a nose.

Teddy bear salad

You can use the ingredients shown below, or anything else you have that can make up a face.

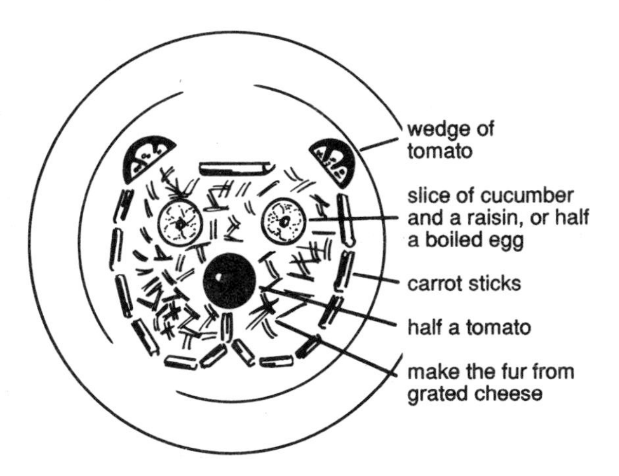

Celery boats

Put cream cheese, or cottage cheese, into bits of celery and add a paper mast on a cocktail stick. Can we make enough so that there will be five each?

Real lemonade

Wash and then cut a thin-skinned lemon into quarters. Put these in a liquidiser with a litre of water and two tablespoons of sugar (or an equivalent amount of sugar substitute). Liquidise for the count of ten seconds – no more otherwise it will be too bitter. Put through a sieve, add more water to taste, and some ice cubes. It makes about 1 litre.

Fruit cup

Put a little of all the kinds of juice or squash you have in a jug. Add water and ice cubes. Slice up an apple and an orange and float them in the drink. If you have some in the garden, you can add a sprig of mint.

You don't need to buy children's cookery books. I used to borrow them from the library and my children never needed to be encouraged to cook when they were little. You could make an exercise book into a recipe book and write in the recipes that your child knows. It is worth buying a good hardbacked book as it may be in use for many years. As your child gets more and more independent he can be encouraged to select something to cook and to do it on his own.

More kitchen maths

Shape and space (what we used to call geometry) is easily learnt in the kitchen.

- Let your child take apart some empty boxes to see how they were made. When the box is flat on the table you can see the 'net' of the shape.
- Then see if the box can be rebuilt and stuck back together again.
- See if they can copy the net with Polydrons (or Galt Jovo).

Capacity and volume are quite difficult unless your child gets a great deal of practical experience.

- Ask him which size of pot would be best for the left-over casserole. (Sometimes it is hard for *us* to get the right size.)
- Let your child pour and fill as much as she wants. They will get some work with sand and water at school, but nothing is quite as good as doing it with you there to talk it through with them.

Questions parents ask

‘ *Should I give my child a calculator?* ’

Giving your child a calculator for their fifth birthday would be a gift with great potential. Teachers who give their children calculators between the ages of five and seven are convinced that those children learn maths much more effectively and much more quickly than other children. I have certainly found that to be true, but I do understand why people are anxious – calculators can be used badly.

‘ *Will a calculator make my child lazy?* ’

No. The reverse seems to be true. Children who use calculators from the early years have a deep understanding and ‘feel’ for number. Giving children open access to them seems to make children think in much more mathematical ways. I think, from what I have observed in my classes, that this might be because when children are freed from the worry of ‘getting the right answer’, they play around with numbers and use their own methods (that they had before they came to school, see chapter 4) and this involves children in very much more maths than we would have covered at this age. For example, I found that children who use calculators a great deal understand place value (one of the most crucial areas of arithmetic, see below) much better than other children.

‘ *Will they be able to add up?* ’

Sensible teachers always encourage children to estimate their answer first. If it is 120 divided by 25, the answer must be somewhere close to 5 because there are four lots of 25 in a hundred, and then there is almost another 25. So if my answer is not close to 5, I know I have pressed the wrong button.

The outcome of all the estimating and all the mental maths that children have to do to use a calculator is that children can actually add up very well. They learn to deal with larger numbers than children without calculators and this seems to make them quite confident at an early age.

‘ *Why do teachers use calculators?* ’

They make children so enthusiastic about maths! Freed from the drudgery that I remember at school of plodding through pages and pages of boring examples, maths with a calculator is such fun that I often had to insist that children stop doing their maths in my classes, when they needed to work at other things. This enthusiasm and positive attitude to maths invariably means that learning is going on, and that learning will be remembered.

‘ *How will a calculator help my child?* ’

As well as being more confident with maths, a child using a calculator is learning to deal with machines that will invariably dominate their lives in the 21st century. It doesn't matter where we go in the Western world, the silicon chip is there. They are in our washing machines, videos, in the cash register at the shop, controlling the aircraft as they land, in the microwave and, of course, in computers and children's games.

Punching numbers into a machine needs to be thought about, not just done mindlessly. Encourage your child to ask each time 'is that a reasonable answer?'.

‘ *Which calculator should I buy?* ’

It doesn't matter at this stage what you buy as long as your child can hold it in one hand and use the buttons with the other. It needs to have big enough buttons so that the wrong one is not pressed by mistake. I like the sort which are solar-powered because batteries are expensive to replace. As your children sometimes drop calculators, my advice would be to go for the cheapest – they are under £5. Once your child is at secondary school she will need a 'scientific' calculator, but these tend to be a bit more expensive and have many confusing buttons.

> *Does it help to buy those supermarket maths workbooks, or will that work against what they are doing in school?*

It can do your child no harm provided you keep it as something that they want to do, or something that you sit and do with them after you have been reading with them, or something that you can use to boost their self-esteem about how well they are doing. They are useful if your child has to miss school for any reason (but remember that a sick child needs something that isn't demanding). What would be harmful is if you say to the child 'they never teach you anything at that school'.

> *My child's teacher talks about "place value". What does that mean?*

Place value is about the way in our number system that a 3 has a different value depending on which column we write it in. So in this example below, we know it means 3 units add 30.

```
 3 +
30
```

We would have no difficulty in working that out to be 33, but a young child might write

```
3  +
30
```

which might give the answer 60. This sort of error is quite common.

So, in number, it matters:

- which column you put a number in (and where you put a

zero, so 03 is quite different from 30, and different from 0.3);

- how you group your numbers. We usually group in tens, but in telling the time we have 60 seconds in a minute, and in imperial weight we have 16 ounces in a pound. We say that we usually work in 'base 10'. This is probably because we have 10 fingers and it gives us the columns, as in this diagram:

tens of thousands	*thousands*	*hundreds*	*tens*	*units*
10^4	10^3	10^2	10	
$10 \times 10 \times 10 \times 10$	$10 \times 10 \times 10$	10×10	10	

This is one of the fundamental ideas about our number system.

When your child is older, he will need a secure understanding of the columns above, and these below that include the decimal point.

· ***decimal point***

tens	*units*	·	*tenths*	*hundredths*	*thousandths*

Your child may play place value games in different bases, for example, in the house game, four windows make a house, four houses make a street and four streets make a village.

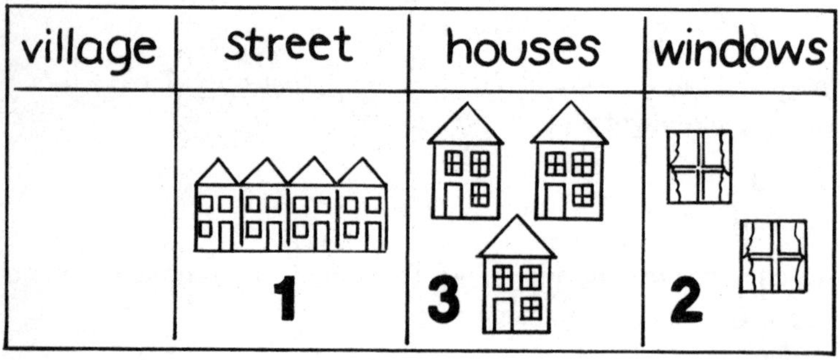

village	street	houses	windows
	1	3	2

The house game

This type of game teaches your child that it matters which column he puts things in and it teaches the crucial concept of **exchange**.

When the child has won four windows in the house game, he can exchange those four windows for a house.

This is the same as – when they are playing with coins – they would **exchange** ten 1p coins for a 10p piece. What is important about exchange is that we are saying these ten 1p coins **have the same value as** the 10p coin. That's a really difficult concept! Try convincing a five-year-old that just because they have a 5p and Harry has five pennies it doesn't mean that Harry has more money than them. ('But he has five coins and I only have one!)

You could make a money game where you throw a dice and win pennies. When you have ten, you have to exchange them for a 10p coin. The winner is the first to get a pound.

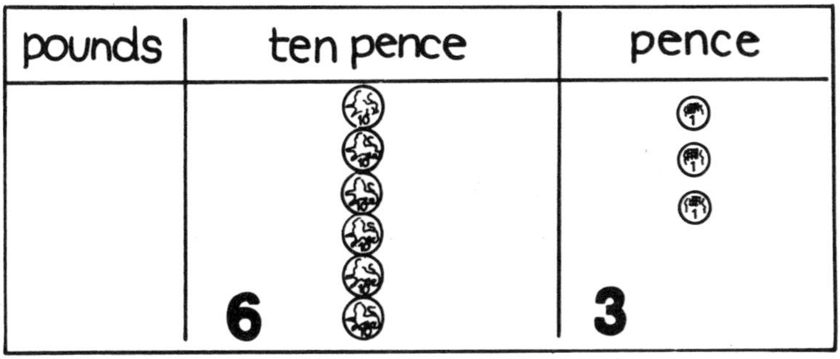

pounds	ten pence	pence
	6	3

The money game

In mental maths your child's teacher will ask children, for example, how the number 34 is made up. It is made up of 3 lots of 10 and 4 units. He will ask how many zeros there are in a thousand and how to write the number that is one bigger than a hundred. (Many children want to write a hundred and one as 1001.)

Helping with place value

One of the interesting things that I find with children who have lots of calculator experience is that they are good at place value.

Here are a couple of widely-used calculator games. (Children can get a great deal from both of these games right up to age 11.)

Space invaders

Take it in turns to key in a three-digit number onto a calculator, such as 369. Give the calculator to another person who has to shoot down the digits one at a time. They can only use one number button each time, but they can use zero as often as they want.

So, to shoot down 369, you could start with the 3. You need to put in −300.

To shoot down the 6 you need to put in −60 and to shoot down the 9, you just need −9.

You should be able to get rid of a three-digit number in three goes. When your child is confident with this, try a four-digit number, then a five-digit one.

Beat the calculator

Write some sums on bits of paper at the level for your child (something like 4 + 3 for a five-year-old and 63 + 7 for a seven-year-old).

The game is for one person to have the calculator and the other to have an abacus, bricks or raisins. They take one bit of paper at a time that has a sum on it and race to get the answer. The winner wins one point or brick.

Some children are very slow with the calculator at first, and they learn the valuable lesson that they can sometimes work it out in their head much more quickly!

Other ways to help with place value are to:

- look at the numbers on the petrol pump or on a digital clock. Notice how the numbers on the right are the quickest to change;

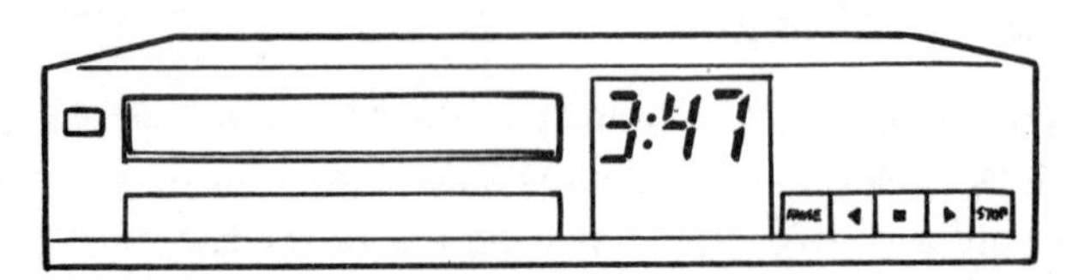

- reinforce left and right. Units are always on the right and many children get muddled by this. Help a child who has left/right and other orientation problems by putting a Red band on their Right wrist.

Many children have problems with left and right and it is particularly common amongst left-handed children and children who have any problems associated with dyslexia. For more detail on this see *Your Child with Special Needs*, another book in this series. Talk to your child's teacher about your child having a red band on sometimes at school, e.g. when she is working with the floor robot and needs to be able to program it to turn left.

The importance of construction toys

In order for your child to appreciate the maths of shape and space he needs considerable experience in constructing things. This can be with bits of 'junk' and an efficient glue, or it can be with construction kits such as Mottik. This is a crucial part of your child's education and if he is reluctant to use construction toys (some parents tell me their daughters can be) then working with your child may get them more involved.

Once children are about five (or much earlier if their manipulative skills are good), they can use Mottik to create a huge variety of models, including those with gears and wheels and with moving parts. Mottik is an exciting new construction toy which won the 1993 Best Toy Award in the construction category, and was a Gold Award Winner in the Good Toy Guide. I like it because it is fun to use, is attractive to both boys and girls and because it can be used more flexibly than some other construction toys. There are triangular pieces as well as cubes, and joining pieces mean that the arms of the dinosaur or the blades of the helicopter move. (You can get Mottik at Woolworths, toy shops, or from the Galt catalogue.)

A Mottik dinosaur

Activities outside

Maths in the park

- Which shapes can we see in the climbing frame?
- Shall we see how many of your paces it is around the pond?
- Can we collect five different kinds of leaves?
- How far do you think it is all around the park?
- How long does it take you to go up and then down the slide? Can you time it on your watch?
- We have just ten more minutes before we must go home. What can you do in ten minutes?
- What do you think is the tallest thing in the park?
- Let's estimate how many people there are in the park today.
- Can you show me how to get home?

Maths at the zoo

The purpose of going on outings is to have fun, so don't spoil it with insisting that they learn something! Make it fun and they will learn.

- You could give your child a plan of the zoo and see if she can guide you around all the animals. (Map reading is important in maths and geography.)
- You could decide to look out for certain things, such as the tallest animal, the heaviest one, the youngest one, the rarest one.
- Look for some shapes, e.g. a circle, a triangle and cubes. (Shapes of cages can be very varied and interesting.)
- Give your child some money and let him decide how he will spend it. Most zoos and museums have pocket money toys.
- Sort the animals into groups as you go round the zoo, e.g. those that like to be awake at night, those that live in water, those that give birth to live babies (as opposed to eggs), or sort them by the number of legs that they have. A seven-year-old might like to learn long names such as 'this is a group of herbivores'.
- When you get home you could start 'a book of outings' and stick the zoo tickets and any other things you collected into the book. You could be the scribe and write for your child if more than a few words or a sentence is too much for her. Don't wreck an enjoyable experience! Your child might want to do some drawings (maybe of the tallest animal she saw) or make her own map of the zoo, or you could help her to make some sets of animals to stick in the book.

If there is more than one child in the family, you will want to find activities that they can all do together. There are ideas for this in the next chapter.

A set of birds and a set of amphibians

CHAPTER SEVEN

Family maths

For those families with more than one child, the times when we can do something with just one of them are so few and far between that it makes sense to build up a repertoire of things that several children can do together.

Family maths at home

Drawing and colouring

Equipment

Drawing involves a considerable amount of maths if you provide your children with everyday stationery and some simple equipment.

A maths set from a high street stationers
A maths set is a piece of essential equipment. The compasses are obviously not for the under sixes, but children need to learn to handle them sensibly.

- Who can make a clown picture just with circles (including drawing around tins and coins)?
- Let's make a pattern just using triangles, or rulers, or a set square.

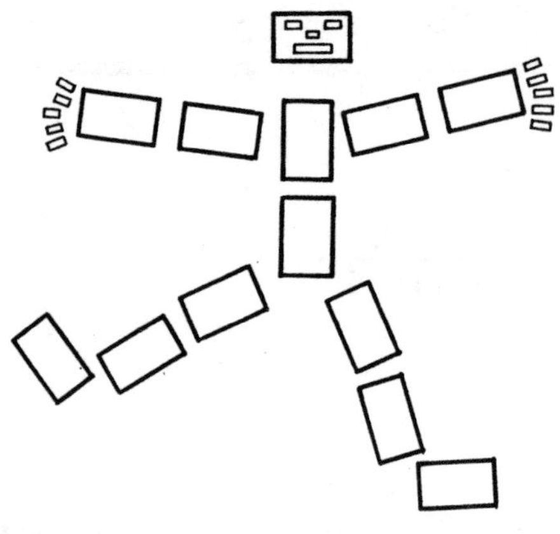

A drawing made just with rectangles

Stencils of any kind require careful estimations and measuring

Those that have geometric shapes (triangles, hexagons, etc.) on them are particularly useful. (You can get these through the Galt catalogue, from Galt Educational, Brookfield Road, Cheadle, Cheshire, SK8 2AL 061 428 8511.)

Pattern pads can be used at any age from about four or five
(One make is 'Altair pads', or you can find 'Hazel Mill' pads in supermarkets.)

- Each child could have a copy of the same pattern and then decide on three or four colours they will use. These patterns often do look best if the children stick to a fairly limited set of colours.
- If you have a large pack of felt tips, one child can have all the blues, one all the greens and another all the reds.
- Who can make a funny looking animal on a pattern pad?
- Who can make a symmetrical pattern?

A pattern has reflective symmetry if one half is exactly the same as the other half and children can test for reflective symmetry by holding a small mirror across the pattern and looking at the reflection. (See chapter 11, pages 167, 170.)

Many children get huge enjoyment from playing with mirrors and there are a number of 'magic mirror' books available from Tarquin Publications and elsewhere. Tarquin will send a catalogue if you write to it at Stadbroke, Diss, Norfolk, IP21 5JP. It produces some of the most unusual, enjoyable and interesting maths publications around.

Cutting and sticking

Cutting and sticking is great fun from about 18 months. Children's glue that will wash out is available. If you cover the table in newspaper, cutting and sticking can be not only mathematical, but also help with the important skills of cutting out accurately and sticking things down exactly where you want them. Collect coloured paper from packets and gift wrapping paper and make a scrap paper box for cutting out. Fabric can be used too, but you might need to provide a good pair of scissors. Remember to buy left-handed scissors (or ones designed to be used by both left- and right-handers) for left-handed children!

Maths from tangrams

Tangrams are fascinating Chinese puzzles, which can be obtained from Tarquin in wood (see the box above). Or you can get an older child to construct some on paper. It's harder than it looks and you might need to help the under-11s.

Once each tangram has been designed, it is cut up and rearranged to make patterns and pictures. If every piece is used each time, with no overlapping, each picture has the same area – the area of the original square. (So if you used a 10 cm by 10 cm square to start with, each picture has an area of 100 square cms.)

On a wet day you could challenge children to come up with 30

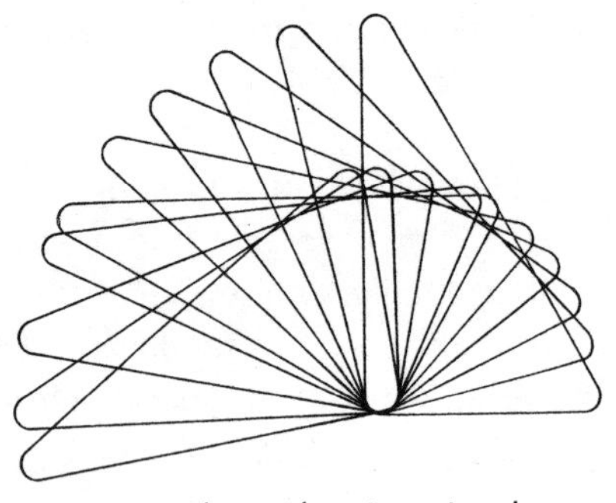

You can make this pattern with overlapping triangles

A Chinese tangram

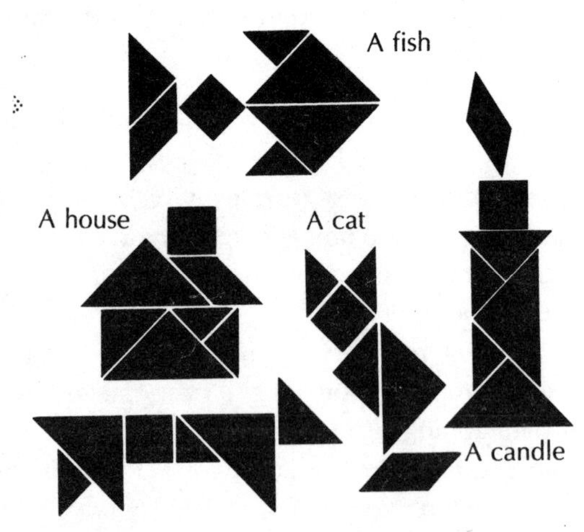

different animals made with tangrams. It can be done if you allow several versions of fish, dogs and cats, etc.

Printing

Printing is much more fun when more than one person is doing it.

It is suitable from about 18 months and projects to do together could include:

- making Christmas/Easter cards;
- sending some pictures to Gran;
- making some birthday cards for friends' birthdays;
- making wrapping paper (you need fairly big bits of paper and it is best if it isn't too thick; if you have a printing firm or a paper mill near you, make friends with it!);
- making games to play (you can make a very satisfying track game if you have car or animal rubber stamps).

- Printing can be done with anything from Mottik to empty cotton reels. Vegetables and fruits are good too. Look for the five-pointed star in apples.
- Encourage designs such as chess-board, or rows such as one of carrot prints and one of lemon prints alternating.
- Round about the age of five children can learn to make simple repeated patterns (see chapter 5 for the beginning of algebra).
- Galt has a good range of printing sets.

Playing indoor games

An enormous number of indoor games is available and many have mathematical ideas in them.

Strategy games

Strategy games often require just two players, but children can play in pairs, or where there is an odd number of children, they take it in turns to sit out, or play a game for one person. Thinking out a move develops mathematical thinking of a very particular kind – children need to learn how to plan and predict.

- Noughts and crosses is great in the car or anywhere. Try it on the usual three-by-three grid, or try it on a four-by-four grid.
- Connect 4 can be played on any kind of grid, including one you make yourself (see page 125). There is a manufactured mini-game for holidays.
- Chess, draughts, Halma and Othello are obvious examples of games to buy, but here are some others.

Frogs (a game for one person)
You put three green counters on each square at the left-hand end of the line of squares and three yellow counters on each square at the other end.

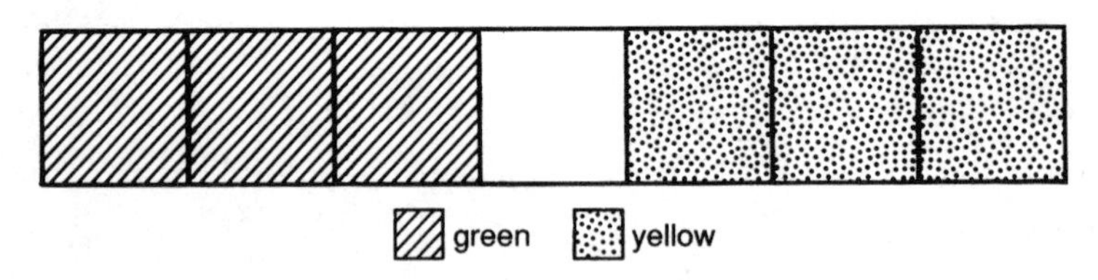

The idea is to swap the position of the green and yellow counters so that they finish up at the opposite end to where they started, but you can only move green counters to the right and yellow counters to the left – no going back.

A move consists either of sliding to the next square, or jumping over just one counter to free one on the other side. No counter can move beyond the seven spaces. It seems impossible at first, but it really can be done.

Your children might have this game on their computer at school. Challenge them to do it with nine squares and four counters each end, or 11 squares and five counters each end.

Three men's Morris (for two players)

You might have seen games like this one scratched on cloister seats in cathedrals. It was a favourite for monks.

You need three counters or buttons each and you need to be able to tell which ones are yours. You take turns to put down a button on any circle on the board.

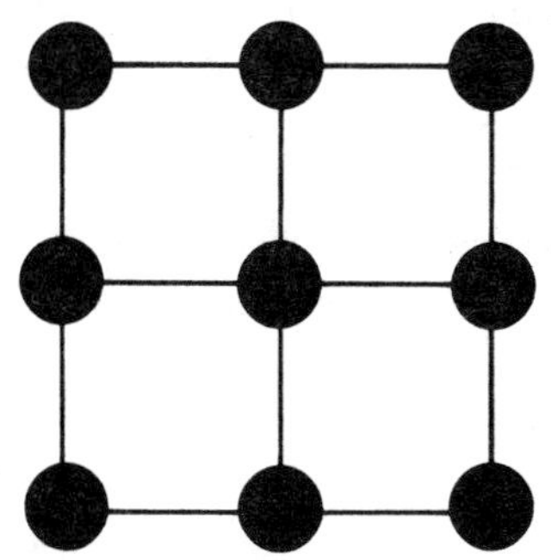

The aim is to make three buttons in a row in any direction along the marked lines (so you can't have diagonals). If, when all the buttons are down, neither of you has made a row of three, you take it in turns to move one of your buttons along a line to one of the empty circles. The first person to make a row of three is the winner.

Nine men's Morris (for two players)

You need nine buttons each for this game – a different colour or size each so you can tell them apart. You take turns to put your buttons on the circles, only one at a time and if you make a row of three, you can take any one of your partner's buttons off the board.

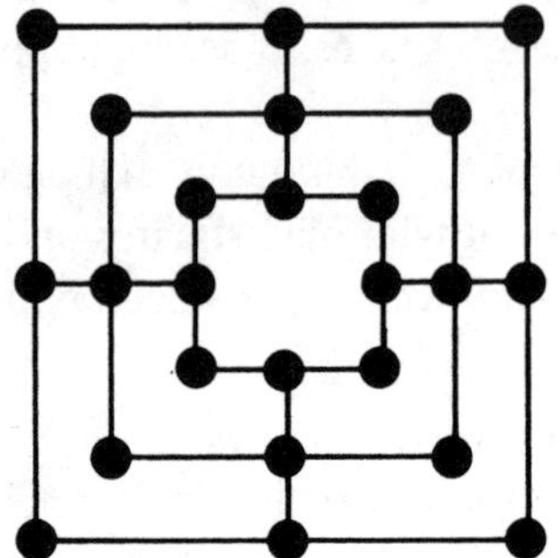

Again, once all the buttons are on the board, you take turns to move your own counters, one at a time, along the lines to free a circle. You have to try to make a row of three and stop your partner making a row of three. As before, each time you make a row of three, you take one of your partner's buttons. The loser is the person who ends up with only two buttons.

Square plan (for two people)

You need 16 buttons to play this and you start by putting a button in each square.

The object of the game is to make your partner take the last button.

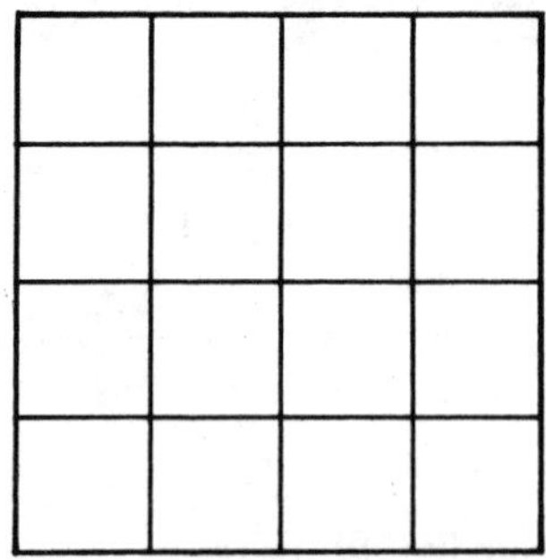

Players take it in turns to take as many buttons as they want from any one row or column. (Rows go across and columns go down.) But you cannot take buttons if there is a gap between them in any row or column. You can only take those that are next to each other.

Forty-seven (for two players)

Young children can use a calculator. You need one button, pencils and paper.

The object of this game is to get a total of exactly 47, or to force the other player to go over 47.

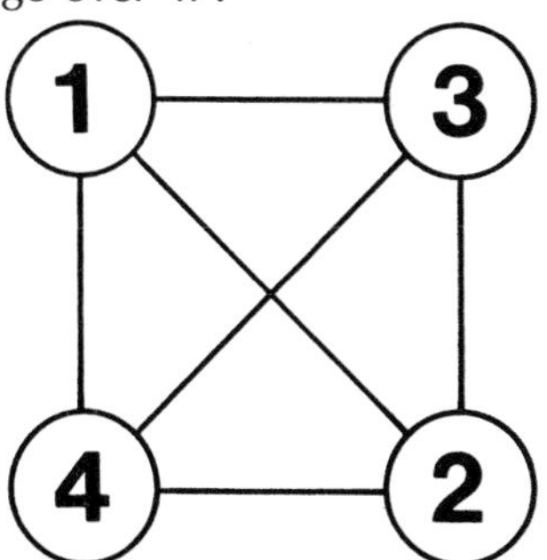

The first player puts a button on one of the numbers and writes the number on the paper.

The second player moves the button to another number and adds it to the number on the paper, putting down the total.

Players take it in turns like this until the total of 47 is reached, or someone is forced to go over 47.

Older children might be able to devise a way of winning every time.

You can play the same game with other totals. Will their way of winning work for other numbers?

Spiral nim (for two players)

There are many variations of the game of nim around, but this one is played on a spiral. (It's still a spiral even though it has square corners.)

- Start with a button on 25. The object is to make your partner move the button onto square 1. Each player takes it in turns to move the button one, two, three or four squares (they can choose).

25 start	24	23	22	21
10	9	8	7	20
11	2	1 home	6	19
12	3	4	5	18
13	14	15	16	17

- You can vary the game and start on square 13.
- This version is quite hard. You play with two buttons on square 25 and take turns as before. You can move either counter, but one counter cannot overtake the other.

  The loser is the one who puts the last button on 1. Not as easy as it sounds!

Dice games

If you have Polydrons at home, or if older children have used the Tarquin 'Make Shapes' books, you could encourage them to make their own dice. Easy six-sided dice can be made from sugar cubes, or from baby bricks. These games are for six-sided dice, but you can adapt a game for any size of dice.

- Throw your number. Each player chooses a number (1 to 6 with one dice, 2 to 12 with two dice, etc.). You can play with as many dice as you want and the idea is to take turns to throw all the dice and to see who can throw their number the most times in two minutes. With two dice this is a very good game to teach quick adding.
- Throwing sixes. Each player needs a 1 to 6 dice. You see who can throw the most sixes in a minute. One person needs to be time keeper and judge.

- Doubles. Each player needs two dice. The aim is to see who can throw the most doubles in two minutes.
- Make a board with the numbers 1 to 6 on it. (For small children make the 'numbers' just like the dots on the dice.) Each person needs a bit of paper to keep her score and you need one cardboard tube – the longer the better. (The middle of a cooking foil or kitchen towel roll is ideal, but you might want a young child to use just a Smartie tube.)

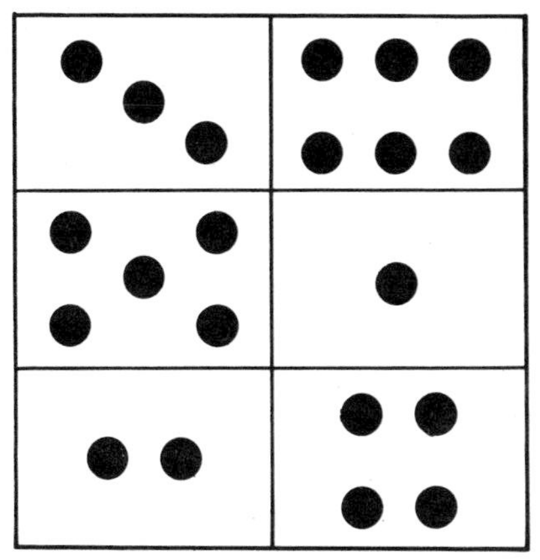

The idea is to roll a dice down the tube onto the board and to get the number on the board and the number on the dice to match. It's harder than it looks. The winner is the person who is first to score all the numbers 1 to 6.

Children can make their own maths games

Children love to make their own games. If you provide them with paper, pictures, stickers and stencils, they will come up with surprising and interesting results.

It helps to:

- encourage them to make the boxes for track games quite large so that their writing fits in:

The monster has got you, go back to the start.

- give them something rectangular to draw around for these boxes (an Oxo packet is about the right size, or a large shape on a stencil);
- give them different dice to play with – get some with ten or 12 faces from a good toy shop;
- show them how to make simple spinners (see page 45) – they need to draw and cut accurately and put the pencil through the exact middle, or the spinner will be biased.

Card games

You probably know many card games and there are a number of books available. The best one for maths that I know which uses ordinary playing cards is *Cards on the table – 20 mathematical card games* by Fran Mosley, available from the BEAM Project (Barnsbury Complex, Offord Road, London N1 1QH). It really is a terrific book for children from about seven. It is best if you play too at first to get the children started.

Kim's game

- Provide a tray with a large collection of interesting bits and pieces – a baby's bottle, packets of herbs, a wooden spoon, etc. and let the children select some other things from places you can decide on (with the rule that what they use they must return at the end of the game).
- Children take it in turns to put a few things on the tray (up to about 20). The others have 30 seconds (timed on a wrist-

watch) to memorise all the things on the tray. The tray is removed and put behind a chair or somewhere out of sight and children take it in turns to say one thing that was on the tray. The checker stays behind the chair and takes things off the tray as they are remembered. The one who can remember the most things on the tray is the winner. Then it is someone else's turn and they choose some different things. One good thing about this game, like pairs (Pelmanism), is that the younger ones can be very good at it.

- A variation on Kim's game is to collect together things that all start with the same sound – glove, goggles, grass, glasses, a green marble, a green hair band, etc. (You need to get some agreement on whether the packet of ginger can be included as a younger child may say that is a different sound.)

Making dens

Can we make a den big enough for all of us to get in and with space for us to have an indoor picnic?

Provide sheets and blankets and let the children move the tables and chairs. Working out space and distance is good for estimating skills – and it keeps them quiet for hours! It's great for when you are spring-cleaning because you were going to move the armchairs anyway.

Playing with the button boxes

Once you can be sure that children won't swallow them, buttons are great to play with. They are used widely in infant schools for counting, sorting and matching, but they are fun for older children too. Children can:

- find the biggest/smallest/shiniest/the ones that I had when I was a baby;
- make a line of buttons right across the table, starting with the smallest button and ending with the largest;

- put them into sets – a set of leather buttons, a set with four holes, a set of shirt buttons;
- glue them onto pictures to put on the bedroom wall;
- make a home-made target and play tiddlywinks;
- glue them onto a postcard with a simple shape to fill in.

They are good too when a child is poorly. Put the box on a tray with something to sort them into (yoghurt pots or a cutlery tray) and some glue and card.

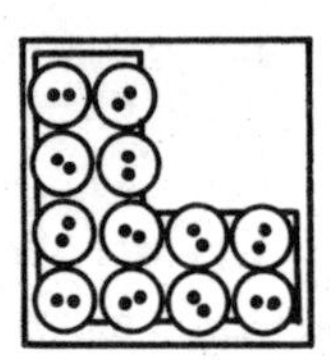

L for Lucy

More games with buttons

Tiddlywinks

You don't need to buy a special board for tiddlywinks. Once children can flip a button, they can create any kind of target.

Suggest that they use an egg box or a bun tray and have bits of paper that show different scores for different 'holes'. (It might help to flip the buttons from a pile of books as a younger child might find it hard to get the buttons high enough to get them into the holes.)

Children can keep a total of their scores.

Button race

This game is a bit hard on the knees, but is great fun. The idea is to flip a button along a track to see who gets to their marker first.

Measure out a track that will fit in the living room or hall but that has different distances for the older children. Clare maybe has to go six lengths of the tape measure, Ellie has to go five lengths and Lucy only has to go four lengths. Markers can be chalk lines if you are outside, or bits of scrap wool for indoors.

Button puzzles (answers on page 199)

1

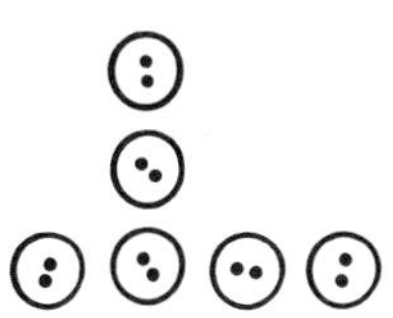

Put six buttons like this and move only one of them to make two rows with four counters in each.

2

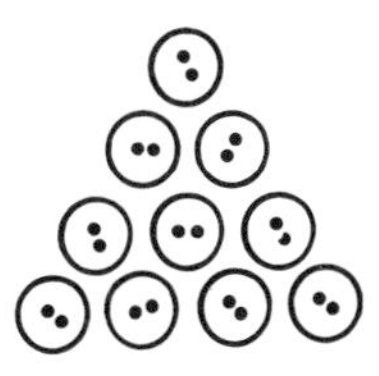

Put nine buttons like this and turn the triangle upside-down – but you are only allowed to move three buttons.

3 Make more triangles with buttons, like this (see over).

Can you see the pattern? What will the next pattern look like? (This pattern gives you the 'triangular numbers'. Ask your children if they can predict what the seventh pattern, or even the hundredth pattern would look like.)

4 Can you balance a button on the edge of a piece of paper?

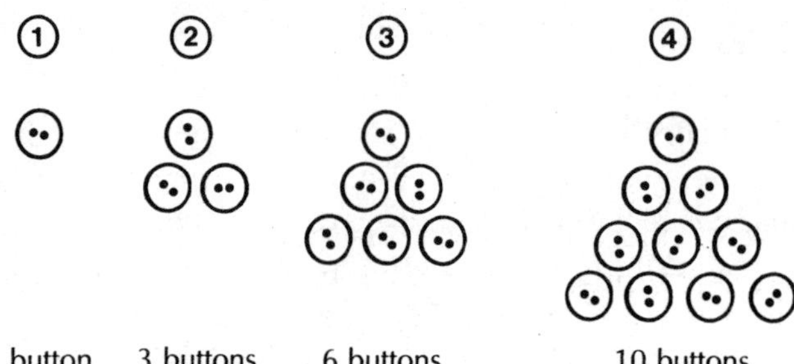

1 button 3 buttons 6 buttons 10 buttons

Weighing

You needn't restrict weighing to just times when children are cooking. If you use the bathroom scales, as well as the kitchen ones, anything can be weighed by a group of children. They could empty a kitchen cupboard and try to guess the heaviest thing in it. (Persuade the children to wipe out the cupboard before the things are put back!)

Food is good for weighing because the weight on the scales can be matched to the weight on the packet. Get children to guess where the dial will come to first.

Construction

Get the children to plan something to make together. Their idea could be based on a story or just their own imagination. What I found helped this kind of play was providing some kind of base board for their work. A large flattened cardboard box will do: they can make it into a roadway, the surface of the moon or a harbour. We had some pieces of hardboard which were just that bit more stable than cardboard. The models (dinosaurs, houses, moon buggies, etc.) are then made using anything available from the train set to bricks and Lego or Mottik. (Take care with play dough though because it blocks up the joins in construction bricks.) By adding some playpeople or peg dolls a small world is created that can last for several days. It can be pushed to the side of the room or under the sofa on the base board. And all those attempts to make things about the right size will help children with ideas about scale.

Treasure hunts

Treasure hunts are enormous fun for children. You can do one with simple clues ('look under the brown time piece') and these clues might give the place of other clues and eventually they find the treasure. Or you could make a simple treasure map ('the treasure is five paces north from the front door step'). Your 'treasure' might be those tiny Easter eggs, or nuts, or those chocolate gold coins.

An older child might be persuaded to make up a treasure hunt for the younger ones. It's a great game to play at a party.

There are many other indoor activities that can help your children with maths, such as dressing up (this dress is too long for me, my feet are too big for these shoes) or counting the money in the piggy bank, that children can do together. I hope this has given you some ideas to get going. Many of the activities can leave you free to do something else, but as with most things, if we join in sometimes, or talk to the children about what they are doing, it can make the learning experience more significant and more fun.

Family maths on the move

Car, bus or train journey games

I spy

As well as the more age-specific games in the other chapters, all the children can play an I spy game to last a long journey.

- Each child could have a colour. They have to see how many cars or lorries of that colour they can see before the journey is over.
- Or you could make a list that they all work at together such as:

  5 black cats

2 ladies with pushchairs
1 red tractor
1 post office van
10 black cars
1 man on a bike
20 sign posts
1 police car
a field of sheep, etc.

Singing

Everyone can sing along with a musical times table tape. Although I have stressed the need to understand tables, no harm can come to a little child joining in the singing.

Sing counting rhymes

There were five in the bed
And the little one said,
Roll over, roll over.
So they all rolled over and one fell out.
There were four in the bed and the little one said,
Roll over, roll over

(etc., up to *two in the bed*)

And the little one said,
Roll over, roll over,
So they all rolled over and one fell out.
There was one in the bed and the little one said,
At last, time to stretch my legs.

Encourage children to make up their own number rhymes. Here is one I have started.

One fluffy teddy bear as hungry as can be
Likes to eat his sandwiches beneath a shady tree.

Two fluffy teddy bears sleeping all the day
One wakes the other up, because she wants to play.

Three fluffy teddy bears . . . etc.

Family maths on the beach

- Who can dig the deepest hole?
- Let's make a huge castle with a moat.
- How far away do you think the horizon is?
- Which way do we need to put the windbreak to shelter us?
- How many miles did we come today?
- Let's see who can find the biggest pebble.

A beach treasure hunt

- Who can find a shell with a spiral on it?
- Who can find something to do with 6? (a crab with 6 legs, a flower with 6 petals)
- Who can find a fossil?
- Let's put these pebbles into sets. (All the grey pebbles, all the pebbles smaller than Ellie's hand.)
- Let's go on a rock pool hunt.

If you feel really brave you can bury something and give the children clues to find it. Don't make it something precious though as they may not find it and you might not be able to remember exactly where it is!

Games with pebbles

Many of the strategy games that you can play with buttons can be played on the beach with pebbles (e.g. Three men's Morris, see page 103).

Pick them up

You need about 12 small pebbles each. Take turns to throw one of these pebbles in the air, and before you catch it you must pick up one of your other pebbles. (Young children might need to catch their pebble in the other hand.) If you succeed with one pebble, you then throw one up again and this time try to pick up two pebbles. The winner is the person who is able to pick up the most stones. To make it easier for younger ones, they can have the smallest pebbles. You can keep scores of each 'round'.

Beach bowling

You need one quite large pebble for the 'jack' and three small stones for each player. You draw a starting line and one person throws the jack – not too far. Then everyone takes it in turns to throw their little stones. Direct hits score 10. The first one to 100 wins.

Target

You play this a bit like beach bowls, but instead of having a jack, you make a target of concentric rings (as on page 48 and in chapter 6, page 81) in the sand. The middle might score 50, and other rings might score ten or 20. The winner is the first to 100.

Magic beach squares

You need three pebbles each and you draw a square in the sand like this one.

4	9	2
3	5	7
8	1	6

The aim is to throw your stones into the squares so that the total score is always 15. You might want to let younger children throw their pebbles from closer to the magic squares and to have more than three pebbles.

Activities for wet days at the beach

Making flags

You need lolly sticks or garden canes, felt tips, paper and glue or sellotape. Each child can only use three colours on each flag. How many different flags can each child make?

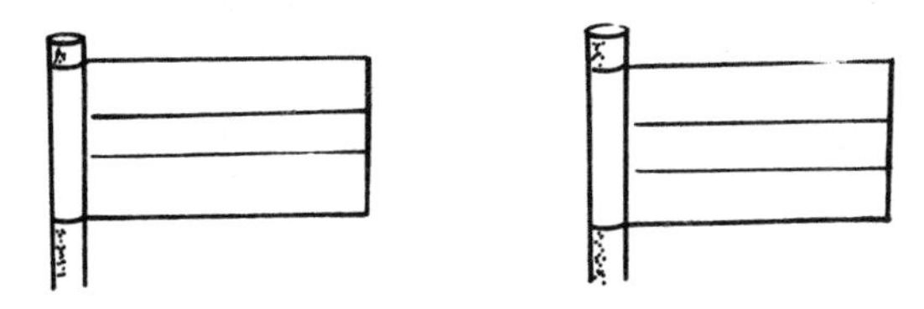

(If they use each colour (red (R), blue (B), green (G)) only once on each flag there are six different ways to combine the three colours: BGR, BRG, GBR, GRB, RBG, RGB.)

Making pebble monsters

Sticking or drawing faces onto pebbles makes an unusual present to take home. See if the younger children can make big monster, middle monster and baby monster.

You can play most of the beach games in either a garden or a park. As a family, you could make a maths trail around your park or town – for details of this, see chapter 9.

Have fun!

CHAPTER EIGHT

Becoming a junior – maths for seven- to nine-year-olds

What to expect at school

When children go into the junior school there is inevitably a change in the ways in which they work. Teachers are encouraging them to become more independent and to work much more on their own and in groups. Just the change to a different teacher is stressful for a child, so we should not be surprised if they react to the change.

If they use a maths scheme at school there is often a very distinct change in the ways that these are organised for juniors. Often there are text books to work from and sometimes this accompanies a change to writing in a maths exercise book with squares. Some children find that very difficult to get used to after the freedom of infant blank pages. At seven, some children still have large writing and the concentration required to get one number in one square on the paper is sometimes too much for them.

Keeping a positive attitude

If your child is experiencing difficulties and is worried about school, have a word with the teacher. He is aware that the change your child has faced is demanding and what you can both do is to work towards keeping the child's attitude towards maths positive.

The main goal is maintaining that positive attitude to maths, and some schools are very good at it. Others are less good and you will need to work hard to counteract that.

What the National Curriculum involves at this stage

At seven your child enters Key Stage 2 and this makes very little difference to the topics covered in maths, except that obviously the work gets a bit more demanding. (For more details about the Key Stages see the appendix on the National Curriculum, page 198.)

The topics in maths are still number, algebra, problem solving, etc. and your child's records of what she has covered and achieved will be sent on from her previous class.

Learning tables

The learning of multiplication tables gets underway in earnest once children reach seven and the junior school. There is not as much chanting of tables as there was years ago, but tables are still taught, though there is a move towards helping children to understand what they are doing rather than just learning by rote.

Learning by rote or learning by heart?

There is a difference between learning by rote and learning by heart. I was taught tables by rote. They had absolutely no meaning to me and if I wanted to know what 7 times 8 was I had to start at 1 times 8 and say the table all the way through. I didn't know that 7 times 8 meant 8 lots of 7 and I'm sure that no-one told me about the connection between multiplication and division.

I have often taught children who have learnt by rote rather than by heart. They are usually confused and panicky about maths, and ask questions such as 'is it a gerzinter?' (Is the problem to do with sharing – a 'goes into'?) Learning by rote does not lead to confident mathematical thinkers. However, if a child learns something by heart that she understands, it can be powerful and useful in her daily life.

What can I do to help?

You can help your child with tables by:

- making sure that he understands what they are all about; and
- helping your child to make connections between things.

What I mean by that is that if he learns the 2 times table, it will help him to know some of the answers for the 4 times and the 6 times and the 8 times. The answers are all even numbers and there are connections you can make between the numbers. It is this kind of 'intelligent' learning that leads to understanding, and then that can lead to learning by heart.

'They need to know their tables to do the shopping'

I am told this repeatedly by parents and by older people who feel that my way of teaching maths is 'all this modern maths stuff'! Of course a child needs to know what the cost of seven packets of sweets at 17 pence each comes to, but she will have a variety of strategies to work it out. I would have been stumped by that as a nine-year-old because we didn't do the 17 times table!

If a child who has been encouraged to think for herself in maths was doing 7×17, she might use her knowledge that 7 lots of 7 is 49. On the other hand, she might do it a different way and work out 7 lots of 20 and subtract 7 lots of 3.

How do I help my child to understand?

It is important to use practical activities to help your child with tables. With bricks or pasta shapes you can put out 3 lots of 4:

and you can rearrange that into 6 sets of 2,

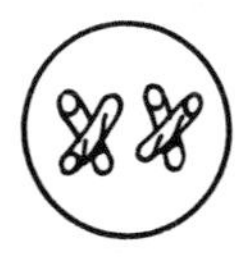

How else can you rearrange them?

From these 12 pieces of pasta you can make 2×6, 6×2, 1×12, 12×1, 3×4 and 4×3. All those numbers are the factors of 12.

This type of activity shows your child the connections between the tables. (So if he knows 6×2 he also knows 3×4.) This makes the job of learning tables much simpler because then children don't see them as dozens of unrelated facts to be learnt.

The 3 times, 6 times and 9 times tables all link together: 8×3 is 24 and 4×6 is 24. The 5 times and the 10 times tables link like this as well, so 6×5 is 30 and so is 3×10. It is helpful to find all these numbers on a table square (see page 122).

Relate tables to everyday life

Do make sure that you relate tables to everyday life.

- If I want about 15 or 20 eggs for the week and they come in boxes of six, how many boxes do we need to buy?
- There are five weeks to Christmas, so how many days is that?
- If I give you 10p for your piggy bank each week, how much will you have after ten weeks?
- What else comes in sets of things? (Biscuits in a packet,

sweets in a tube, buns on a bun tray, tennis balls in a box, etc. . . .)

Other activities

Let your child make her calculator multiply

You can often make the calculator multiply by entering **2 + +** then =. Then if the = is pressed repeatedly, it will go on adding 2, and will display 2, 4, 6, 8, 10 and so on for as long as you want.

You need to find how your calculator will do this. It might be **2 + = = =**, or if you press **2 + +** you may see a little **K** appear on the screen and then pressing = will make the numbers appear.

You can do the same for all the tables and it is helpful if your child learns to count with the calculator, saying 2, 4, 6, 8, etc. out loud. Children can generally count like this before they can say a table by heart.

Make a table square

Your child may make a table square at school. These are used as a quick reference for tables and it might help to make a big one and put it up on the bedroom wall. If your child is by now quite anxious about maths, it would be better to put it up in the kitchen than to invade his place of safety and rest with a reminder of the cause of the anxiety.

X	1	2	3	4	5	6	→
1	1	2	3	4	5	6	
2	2	4	6	8	10	12	
3	3	6	9	12	15	18	
4	4	8	12	16	20	24	
↓							

Lots of ways

Play a game with your child while you are busy making a meal, or while you are walking to school. Choose three related numbers (3, 4 and 12, or 6, 5 and 30) and find as many different ways as you can of expressing those three numbers as multiplication and division facts.

So, for 3, 4 and 12 you might have:

1. 3 lots of 4 makes 12 altogether.
2. 3 towers of 4 equals 12.
3. 4 + 4 + 4 (it is good to show your child that multiplication means multiple addition).
4. 3 + 3 + 3 + 3 = 12.
5. 3 times 4 makes 12.
6. 3 multiplied by 4 makes 12.
7. 4 groups of 3 is 12.
8. 12 divided by 3 is 4.
9. 12 things shared between 4 people is 3 each; and so on.

Keep it fun and let your child be creative about it. This is a good way to help him to remember those 'awkward' table facts like 7 times 8 that many of us find difficult to remember.

Tables tapes

You might find one of those musical times tapes that sings the tables is quite helpful. I know that for some children, their tape really helped them, but your child is unlikely to choose to listen to the tape on her own. Do ask around and see if you can borrow one first. There are several on the market now but some people hate them! A tape might be good for long car journeys, or for playing and singing through together as a meal is being prepared. Don't choose the moment when their favourite television programme is on!

As with any maths that you do with your child, be led by him and follow his wants and needs. Don't add to the anxiety, but focus on building up his confidence and independence. Keep in touch with the teacher and try to be aware of what your child is doing at school.

'My child is confused about division. How can I help?'

Children can get horribly confused with division! I've seen children panic as they realise the task is a 'gerzinter' (it goes into). Here are a few simple activities that you can do to help them.

- Make sure they have understood what multiplication means. It is multiple-adding, so three fours means 4 + 4 + 4. (See above.)
- Make sure that they can draw a picture of three fours. (See above.)
- Make sure they see that multiplication is the reverse of division, so their table square (page 122) can help them with division.
- Play 'the magic three numbers game'.
  1. The child chooses two numbers below 10, say 3 and 4.
  2. The adult has to work out what the third number is (12 in this case: 3 lots of 4).
  3. The child then has to fill in all four columns on a piece of paper.

3×4 = 12	4×3 = 12	12÷3 = 4	12÷4 = 3
5×6 = 30	6×5 = 30	30÷5 = 6	30÷6 = 5

  4. Make sure she can relate this to buttons or bricks and can take 12 pieces and actually make a picture of each

of those equations. (See the pictures above.) You could keep the paper on the front of the fridge and add to the columns whenever there is a moment to play the game.

- Play 'sausages' (see page 156).
- Make real situations for division. Invite friends round and suggest that the biscuits, Smarties, or nuts are shared out evenly.
- When you do tables practice with your child, relate this to division:
  1 Which two divisions can I get from seven eights are 56? (56 ÷ 7 is 8 and 56 ÷ 8 is 7.)
  2 Let children do divisions on a calculator. This will encourage them to be more adventurous and learn to deal with funny remainders.
- Play four in a row (see below).

Playing 'Four in a row'

Here is a suggestion for a 'Four in a row' game (played like Connect 4, the game with counters). This one is based on division (see page 126), but you could make one based on any aspect of number.

I have photocopies of a blank grid and clouds which I use to fill in numbers that will suit the needs of the children I am working with. You can do a really simple adding game for small children and a very complicated multiplication one that needs a calculator.

You need:

- two players;
- counters or buttons in a different colour for each person.

Choose a number from each of the number clouds (you can choose the same number twice) and divide the first by the second to make one of the answers on the playing grid. You then cover that square on the grid with one of your counters. Then the other player has a go. The aim is to get four counters (or three for a quicker game) in

10	8	12	9	7	12
3	2	3	4	8	3
4	5	10	6	5	4
8	3	9	8	4	8
2	7	12	2	5	2
8	4	6	9	7	3

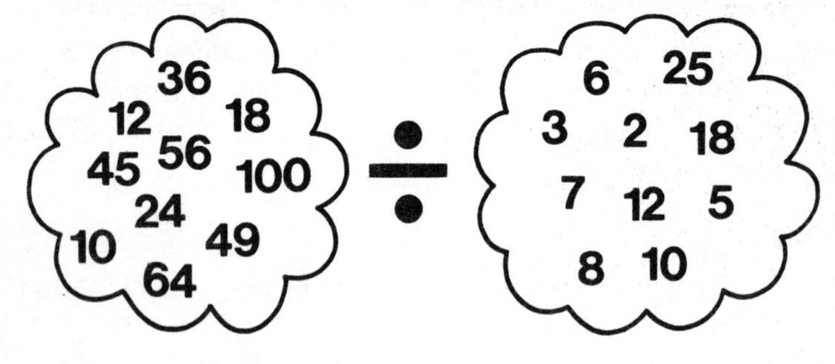

a row, in any direction, up and down, left to right or diagonally. Only one player can have a counter on any one square.

You can see how it becomes a strategy game. If I want to cover the 12, I need to work out which two numbers I need.

'Why can't she tell the time yet?'

Telling the time is one of those things that we think of as easy but is, in fact, really quite difficult. I have had many parents coming to me very worried that their child has not learnt to tell the time by about age six, but this is really nothing to worry about. My son didn't learn until he was almost nine and I know I was concerned about this but when I thought about it, I didn't have a single clock in the house that was easy to read! Once I realised this I bought a cheap round-faced one with all the numbers on it and gave him a watch and the problem was solved in a few weeks.

Of course, the watch was digital and the clock showed traditional analogue time. That's part of our problem now because children find digital very much easier. I wonder if maybe we should teach digital time first? It is the language of timetables and most cheap watches are digital.

If your child is having difficulty with learning time, buy him a watch and set him times to do things by:

- You can watch television until six o'clock.
- Please come back for tea at 5.30.
- You must be in bed by eight o'clock.
- The pizza takes 15 minutes to cook, so can you time that on your watch?

Many digital watches have stopclocks and children love playing with them.

- Time yourself to tidy up your bedroom.

- Please could you tell me when the eggs have boiled for four minutes?

Make sure that he understands about the passage of time. (See page 53.)

- You might talk to him about seasons, days of the week, months, lunar months and things such as leap years.
- If he is interested in sport, link leap years to the Olympic games. Have a clear calendar on the kitchen notice board and draw his attention to things written on it.
- We need to take your library book back tomorrow.
- It's Gran's birthday in two weeks.

Buy your child a Christmas advent calendar; you can do similar countdowns to special treats and birthdays.

Do they really need to learn the 24-hour clock?

Children find this much easier than we do. They need to use the 24-hour clock to be able to read timetables and some digital clocks. My own method for working out 13.00 hours is to take away 2 and ignore the 'tens' figure: 3 take away 2 is 1, so 13.00 hours is 1 o'clock. 17.00 hours is 5 o'clock, because 7 take away 2 is 5.

That's my little trick for myself and I don't teach it to children. I teach them to learn a few times, such as we go home from school at 15.30 (3.30) and they soon learn to work things out for themselves. They might use a trick like mine, or teach themselves some quite different method for remembering. What matters is to be able to work it out so that you don't miss the train.

Checking up that your child understands place value

(See chapter 6, page 89.)

The important concept of place value underpins all number work. You can help with this by giving your child a calculator (and perhaps borrowing a book of calculator activities from the library) and by doing some of the following activities.

- Using a set of nine numbered cards (marked 0 to 9), deal out two cards to each player. Then the cards are turned over and the winner is the person who can make the largest number.

92 is larger than 78

Vary the game as the child gets older by building up to dealing out all the cards, then each player uses her cards to make the largest number possible and tries to say what the number is. The winner is the person who can make the biggest number. You can vary it so that the winner is the person who makes the smallest number using all her cards.

0149 is smaller than 2358

- Can your child add 10, or 100, or 1000 to any number?
- Can your child read large numbers to you correctly? Try with car registration numbers, e.g. 4591 is 'four thousand, five hun-

dred and ninety-one'. Don't let him say 'four, five, nine, one'.

- Does she know which numbers come after 9, 19, 99, 109, 199, 999, 9999 and can she write one million in figures?
- If something in the catalogue is £5.99 or £9.99 can they tell you the price to the nearest pound? (Rounding up or down is a good skill to learn by about seven or eight.)

Home/school maths links

As your child gets older there is a tendency for there to be a bit less contact with the teacher. That is partly because the children get more independent but also because once they are seven, many of us are working out of the home for longer. That makes it harder to help at school and be involved.

The teacher of seven- to nine-year-olds needs help too and if we can offer to play maths games or supervise children on the computer, that is a great help.

I found that I missed the close contact with school that I had before I was working out of the home, but one clear advantage for my children's maths was that when I was working I could afford to buy more books, construction toys and maths games.

Home/school projects continue and develop as children get older

Home/school maths schemes usually go on up to 11 and maths games libraries have many games for older children. If your school doesn't have a home/school maths project, suggest that it thinks of starting one.

You, or the school, could contact IMPACT and it will help you to set up your project. It is in the School of Education, University of North London, 1 Prince of Wales Road, London NW5 3LB.

The sheets your child will bring home will not be 'homework' as we remember it, but something like counting up how much liquid each member of the family drinks over the weekend.

If the school is not interested in home/school maths, or is busy doing something else, make sure that your child has access to a variety of mathematical games and activities.

Patterns of behaviour

With my children, this was the age at which I think that their patterns of home behaviour became more fixed. The one who hated the television came to hate it more and watch it less and the one who would sit forever transfixed by the screen became hooked on computer screens as well.

These patterns remained the same for many years so it is between the ages of seven and nine that family patterns of behaviour, such as playing games, or building models together, become even more important. We can change these patterns of behaviour, and if you have a family of avid screen watchers, I recommend buying a new construction kit, or some mathematical games and sitting down and playing with your child.

Playground games and rhymes

Between the ages of about seven and nine, children have a huge variety of playground games that they play and rhymes that they use to skip and play ball. It really is the age of the game. They love to follow rules or to make up their own. There are playground 'fads' that they all follow like lemmings! It might be yo-yos, marbles, jacks, skipping, cat's cradles or something else that the whole world is doing so they simply must do it too to keep up their 'cool' image.

These traditional games (and many of the new ones too) are rich in mathematical ideas so do encourage them. Look for these games in the better toy shops. They are often cheap to buy or make and you can reminisce about how you used to play cat's cradles and can probably show them how to improve their game.

Strategy games

Strategy games develop logical skills and thought about sequences and forward planning. They are games for young and old, so get some for the family. They include: Othello, chess, draughts, Halma, Solitaire and many others.

Many of these games can be played in the playground or at home with just a few stones or dice and I have put examples of them in chapter 7.

School playgrounds

School playgrounds influence the ways in which our children play and learn, and so they are important places. But some of them are so boring! No wonder there is sometimes bad behaviour – I should think much of it comes from boredom. It is incredibly easy to plan games for the playground. With a spare can of emulsion paint all sorts of shapes and games can be painted on the ground. You could suggest this to the school. Maybe the children could plan what they would like and then you could supervise the painting?

Here are some ideas:

- hopscotch;
- an enormous chess and draughts 'board';
- snakes and ladders;
- a long snake number line, 1 to 100;
- an outline dinosaur that is 50 children long;
- 'boards' for various strategy games;
- a target for games involving throwing balls and beanbags;
- a race track;
- various coloured trails to follow;
- the points of the compass and a treasure map;

- a square, a rectangle, a triangle and a circle that each have an area of about 30 children standing close together.

Ask the children. They are usually very creative.

CHAPTER NINE

Things to do with your seven- to nine-year-old

What your nine-year-old can do

By this stage in the child's school career there is an enormous difference in the abilities of children so it is hard to say exactly what your child will be able to do. However, by this stage, many children can:

- use quite large numbers confidently;
- use and understand multiplication up to about 11 × 11 (children only need to learn up to 10 × 10 but the 11s are fun);
- use a computer and calculator with great confidence;
- solve everyday problems such as working out if they were given the right money as change;
- just about tell the time if they have a watch;
- use the points of a compass and read a simple map;
- play a strategy game such as chess with considerable skill;
- cook independently;
- measure length in centimetres fairly accurately and use other tools for measuring, such as a litre jug and scales;
- use a pair of compasses to draw a circle;
- use Logo on the computer (see page 147);
- tell left from right (unless they have orientation problems);

- know what a right angle is and be developing a sense of sizes of angles in general.

Maths isn't just about number

Maths is now a much broader subject than when we were at school. Nowadays it also covers shape and design, measuring with real measuring instruments (such as a clinometer for measuring heights of trees), handling information and putting it on a computer, and statistics. These other areas of maths can arise naturally in our everyday life so it is important to involve our children in the maths that we do, such as gardening, predicting how much milk we need for the weekend, how much wallpaper we need for their room, and whether we decide to go for the expensive all-in holiday insurance, or just pay a bit and risk it.

Building on hobbies and interests

The seven to nine age group is the exciting time when children develop all kinds of hobbies and interests. Lego, Mottik and Meccano are used in more ambitious projects and other art and craft activities, such as woodwork and printing, can be explored. Children like to try out anything different, from learning a musical instrument (music is very mathematical) to making scale models of spaceships. Of course they are not likely to do these creative things if they spend most of their free time watching the television! It's no good moaning at children and telling them to go and do something more interesting – they still need us to be involved at this stage and to take the lead.

Art and craft

If you or your child are at all artistic – or would like to be – there are a number of things available that can be the basis for exciting mathematical learning.

The Tarquin products are some of my favourites. You can now get them in some high street shops but I recommend that you send for the catalogue anyway. I have used these at home and at school – and as very successful presents for nieces, nephews and godchildren. Write to: Tarquin Publications, Stadbroke, Diss, Norfolk, IP21 5JP.

Here are some good ones for the seven to nine age group:

- Making gift boxes. These are cut out and stuck together to make many different shaped boxes that are ideal to put a small gift in – great if you have a sociable child that seems to be invited to a party every week! (*The Gift Box Book*, Jenkins & Wild)
- Magic mirror books that teach symmetry in an enjoyable way. (*The Mirror Puzzle Book*, M Walter)
- Making moving patterns – a mindboggling book that touches on optical illusions and makes an unusual birthday present. (*Make Moving Patterns*, J Armstrong)
- Books and posters about tessellations (shapes that fit together).
- A book that cuts out to make a globe (suitable from about 9). (*The Tarquin Globe*, Jenkins & Bear)
- A book that cuts out to make a star globe – a great project for winter and to introduce astronomy to a child. (*The Tarquin Star Globe*, Jenkins & Bear)
- Many books on making geometric patterns – my personal favourite is one called *Polysymetrics*. (*Polysymetrics*, J Oliver) It sounds complicated but at under £3 it is a more

creative version of the Altair pattern pads you can buy in stationers.

Using a maths set

Buy your child a maths set that includes a pair of compasses, a set square and a protractor and let her explore the designs that she can make from these. She will need lots of scrap paper and some felt tips.

Flower patterns

You might remember making this flower pattern when you were at school.

This will help your child to get better at drawing circles. Help him by:

- giving him a magazine to go under his piece of paper for the compass point to go into;
- suggesting he turn the paper rather than the compass.

(See chapter 11 for some more examples of patterns from a maths set.)

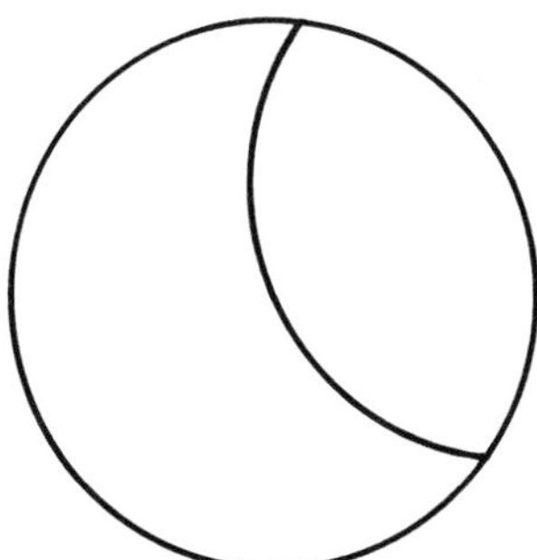

Draw a circle and – being careful to keep the compasses set at the same distance – choose a point on the circle and make an arc with the compass

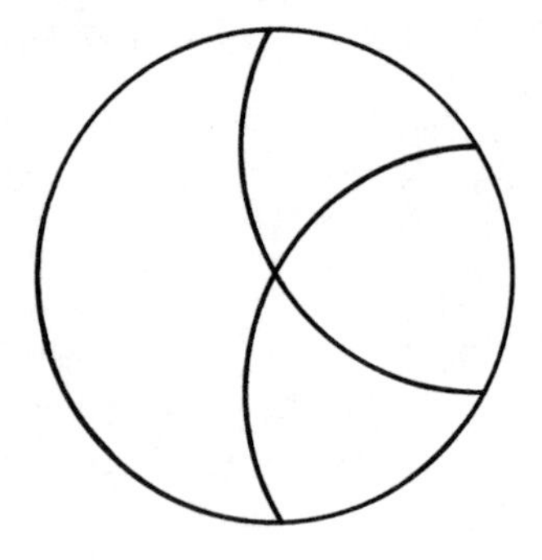

Put the point of the compass where that arc cuts the circle and draw another arc. Put the point of the compass where that new arc cuts the circle and draw another one, and so on, right round the circle.

You will find that you now have a six-pointed flower!

Islamic patterns

On your maths trail (see below) you might find Islamic patterns. These are great to draw on squared paper.

They must be symmetrical. Get your child to test this with a small mirror. Once a symmetrical unit has been drawn, it is repeated across the paper. (Just make a small strip of paper, otherwise it takes ages.)

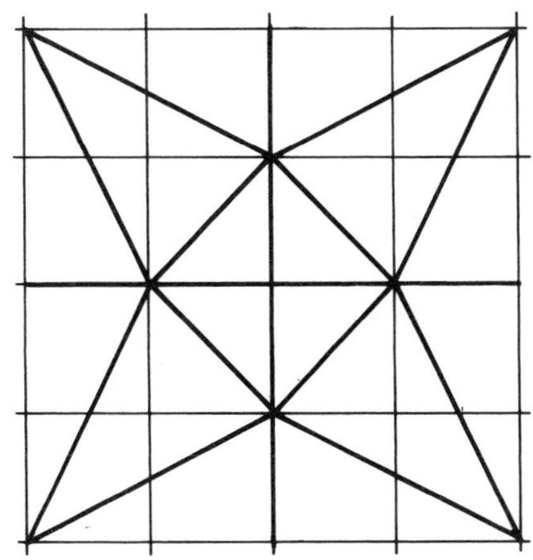

Angles activities

The thing about angles that I never picked up from school was that angle is actually about movement. I thought it was about two lines drawn on paper with a little arc thing between them. I never saw angles as a bit of life. They are there:

- between the hands of a clock;
- when we use scissors;
- as we open the door;
- in patterns and designs.

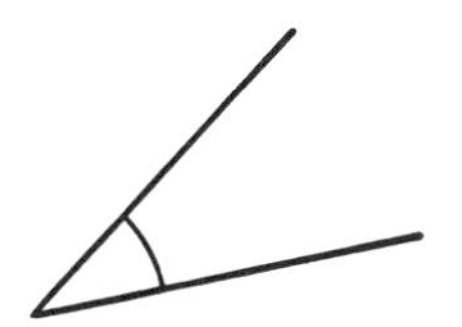

Children use angles a great deal in school now, especially if they use Logo on the computer, so a thorough understanding of angle is important. If your child is a bit confused, make a simple angle-measurer. You need two different-coloured bits of paper and a pair of compasses.

Draw two circles (the same size) on each bit of paper. Cut them out and cut a line into the centre of each of them. Now link the two circles together and put them flat, and they will move around each other showing different sizes of angles. Ask your child to guess 90 degrees, then check it on a protractor. Can they guess 30 degrees? Can they put the angle measurer to the size of the angle between the clock hands at 12.20?

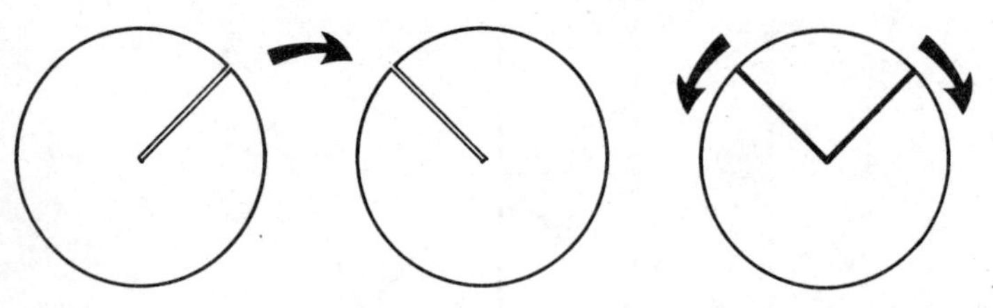

Make sure that your child has a maths set with a protractor and show her how to use it. There are better protractors or angle-measurers in schools sometimes, but I have never seen one in a high street shop.

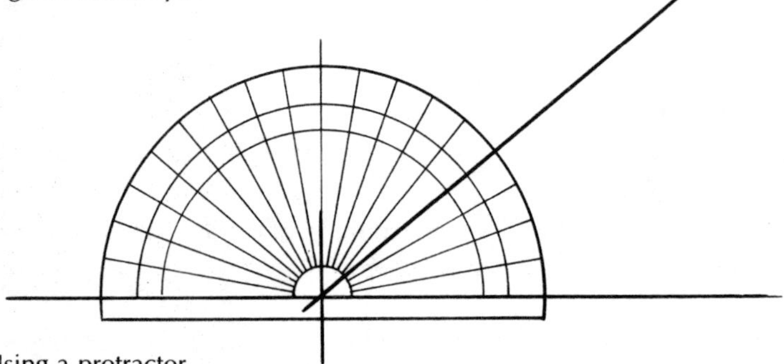

Using a protractor
You put the zero on the protractor on one of the lines to be measured, and its mid-point at the join of the two lines that make the angle. Measure round, making sure you start at the zero

Making cards and patterns

Your child might like to make some birthday or Christmas cards, or he could print some wrapping paper with shapes that tessellate. You can make suggestions such as sticking an origami shape onto card or making a pop-up card. The card shown at the top of the next page has the caption 'Hang in there!' These require accurate measuring and careful folding. Or you can make a picture with one of the ancient tangram puzzles from China (see chapter 7, page 100).

Woodwork

Many children learn woodwork at school but then never have enough time to develop their skills. If you have a woodyard near you, or a DIY shop that cuts wood, you can go and ask for offcuts. I used to take some to school and take some home. Balsa wood is terrific to work with, but it is expensive. Your child needs a few good tools, nails and some kind of old table or workbench. The measuring is obviously mathematical and the constructing touches on technology and science as well.

Of course she will need supervising for sawing and drilling, but if you teach her some basic rules of safety, you will be able to judge what she can do on her own. Good toy shops have child-size woodwork tools, but once children are about eight or nine they prefer to use adult ones.

Learning to measure

Children learn to measure things by seeing us measuring and by actually doing some measuring themselves. If we watch ourselves measure, we will see that although we know how to use tape measures and scales, we often measure more informally. For example, if we work out how wide a piece of furniture is to see if it will fit in a different place, we often use hand spans or some other part of our body to measure it – we do not always do it with the tape measure. When we cook we use spoonfuls and cupfuls and make guesses.

Longer projects

Once children are eight or nine they can become involved in much longer and more complex home maths projects. But our children are products of a culture that is fast, gripping and visual. You and I might have been content for hours with a book, but the child of the video and Nintendo game era is much harder to please – especially if he gets hooked on television or computer games. There is a very real sense in which we may need to learn to entertain our children in a way that our own parents wouldn't have dreamt of.

Orienteering

The best way to get into orienteering is to borrow a children's book on it from the library. Your child will need a compass and a map and a few friends – and a watch so that she gets back at the agreed time. Scout groups are good at this kind of activity so see if there is a local one to join.

Maths trails

A maths trail is a bit like a nature trail at a nature reserve, but instead of looking at ponds and heaths, the children have to look for the mathematics around them in the town, the school grounds, or at the zoo.

You can make a maths trail anywhere, and along your street might be a good place to start. I would think that if you developed some confidence with making a maths trail and then offered to make one in the school grounds, the school would be thrilled.

Here are some starters.

- Look at doorways – are they arched, rectangular, does the brick pattern change around the door, what patterns are there on the door and what shape is the door knocker or the bell?
- Look at brick patterns. Some of them have smart names.

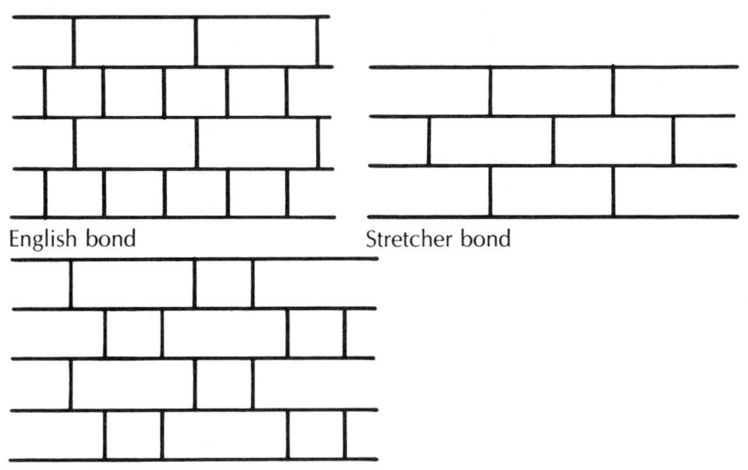

Flemmish bond

You could ask children doing the trail to sketch two different brick patterns.

- Look at bridges and other strong structures such as cranes. Triangles are very strong shapes and these are often seen where a structure needs to be strong.
- Look for tessellations in floors and tiles – which shapes have been used? Your child could try to copy some of the tessellation shapes with Mottik.
- What is the tallest thing in our town?
- What is the oldest thing in the village?
- How tall is the big oak tree? How could we measure it?
- Sketch some roof tile patterns.
- How far is it from the cross-roads to the centre of town? The sign post says $\frac{1}{4}$ of a mile – is it right?
- Can you estimate in acres how big the park is?
- Where does the river go once it goes under the bridge in Church Street?
- Can you find some signs that are symmetrical? (The British Rail symbol, and the Westminster bank sign both have rotational symmetry.) Can you find more examples of symmetry?

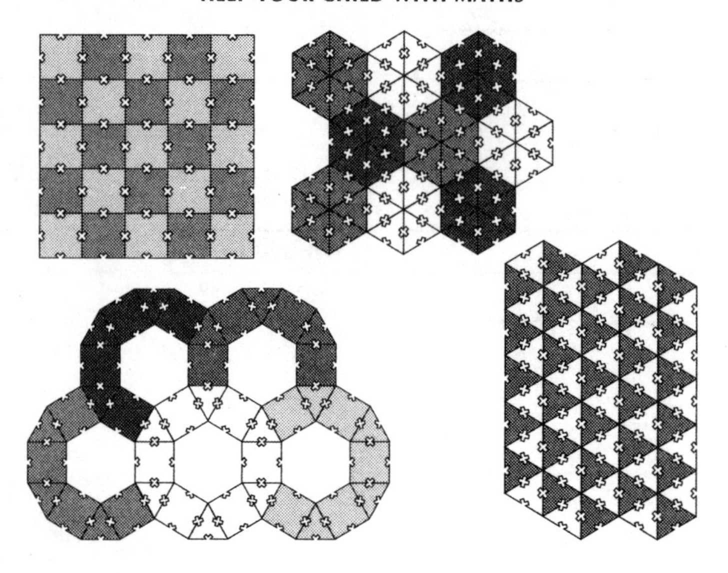

Tessellations made with Mottik

- One of the houses in the High Street has the date 1877 on it. Which house is it?

You can see that whether you live in a city or in the country, making a maths trail is a possibility. If there is a local centre of interest such as an old church, a castle or a cathedral, you might find many dates and a huge variety of shapes. (Some of the shapes in a church vary with the architectural period in which it was built and you can find out about this in a child's book of architecture.)

Encouraging girls

As we have said before, maths is often seen as something that males are much better at than females. There are probably quite complex cultural reasons for this but the issue for us if we have a daughter is that we need to be as encouraging and supportive to girls as we possibly can. (Of course, we must encourage boys too.)

As far as I can understand it, it just isn't true that girls are not as good at maths as boys. They can be culturally conditioned to think that, but there are no reasons to believe it. Sometimes as mothers and teachers we might convey unintentionally that women are not good at it.

- 'We'll ask dad to do that when he gets in.'
- 'Oh, I don't know, ask dad.'

At school, boys tend to dominate the computer and the more mathematical equipment such as the construction equipment. They are given toys like these at home so they are good at using them. While the boys build a racing car with Lego Technic, the girls tidy up and draw houses with flowers around the door!

We can, and we must, challenge these stereotypes for the sake of those girls.

What to do if a child develops a 'maths block'

It may be that your child has difficulties with maths although he can manage other areas of the curriculum. If this happens you will soon become aware that there is very little support given to children like this in some schools. Naturally enough there is often help given to poor readers, but you might be fighting a losing battle to get help with 'maths block'.

In my experience, 'maths block' can be about:

- not getting on with the teacher;
- missing some vital part of the maths curriculum;
- being a slow developer;
- being afraid to be shown to be failing;
- a deep fear of maths that developed from past experiences of failure;
- lack of experience with maths apparatus so the child fails to

develop mental images – he cannot 'see' how to find half of 57;
- having a way of working at maths that is fine when the maths is easy, but breaks down once the work gets harder.

Your child may have developed 'maths block' for many reasons, but whatever they are it can be overcome. Like reading difficulties, if the child is given extra help by a sensitive and caring adult, she can catch up and get back her confidence.

- If you work at the activities in this book with your child, that will help.
- Discuss the problem with your child's teacher.
- Make sure you keep encouraging your child and praise what she does.
- Try to get some help from your primary school because this could help your child to be ready to face secondary school maths in a few years. The earlier you get help the better. There might be other children in the class who would benefit from extra help.
- If the school doesn't have the resources or the expertise to help your child, try to find a tutor (see page 183). Again, if there is another child that needs help, that shares the cost.

Which computer?

The computer will probably be significant in many of our children's adult lives. Having a computer at home can help their learning – but beware, many computer games are just home versions of games arcades at the seaside! Unless you want your child to spend hours shooting down aliens, you need to get some information before you buy.

I'm afraid that when parents ask me which computer to buy, I don't have any easy answers. It would make sense to get the same make as the ones at school, but these can be very expensive.

Cheaper makes in the shops may not have the educational software (the programs you put into the computer to make it do things) that your child uses at school.

Programs to buy

If a computer has all or most of these types of programs, it will help your child to learn.

- A Logo program is the most useful mathematical tool that you could possibly get on a computer. Logo is a computer language that is designed for children. They can draw with it, and produce sophisticated designs (some of which they move around the screen) and will be learning an enormous amount about maths while they are doing it. Logo can be used by children from when they first go to school, and they can still be learning from it at 18.
- Other kinds of drawing program are a good buy. Sometimes these are linked to programs that help the child to make a newspaper.
- Adventure games are enormous fun. These are quite different from arcade games and require a great deal of thought.
- It is a good idea to buy some kind of word-processing package.
- Other 'educational' games may not be very helpful. Find someone who has got the program in question, or ask a teacher, or look in computer magazines in the local library.

Toy robots

Computers in toys are increasingly common. At school your child may use a floor robot or a turtle. Anything that encourages him to do simple programming will help him mathematically.

Don't worry if you can't get a grip on the computer and the robots! Your child will do things with it that will amaze you.

CHAPTER TEN

Maths for the nine- to 11-year-old

What your child can do at this stage

By this stage we hope that children are working confidently with maths. They can more or less use all of the four rules of arithmetic (+, −, × and ÷) but they will not understand all the aspects of these four rules. They know the names and the properties of most 2-D and 3-D shapes, they understand angles, rotation, symmetry and can do elementary algebra such as putting in the missing numbers in these:

4 + □ makes 12

8 + □ + □ = 40

They are quite good at measuring and can make a good go at reading a thermometer as well as other measuring instruments such as spring balances, bathroom scales and tape measures. They can use maths in their everyday life – budgeting the pocket money, getting themselves somewhere on time, and cooking and shopping independently sometimes.

What to expect from school

The curriculum at this stage includes all the kind of maths that we might expect – quite complicated geometry, decimals, fractions and getting into percentages, as well as continuing to focus on practical maths and maths in everyday life. Your child will probably be tested at some stage round about 11 (see the appendix on the National Curriculum), but those tests will only be on the maths that your child does every day and there is nothing for you, or your child, to get anxious about.

How to help

Children at this stage still need to be doing practical maths. If the school believes that children need to do without apparatus by this age, you might need to do some practical work at home to compensate for this. Remember that maths is something that is about what goes on in our heads and you will not necessarily be getting much insight into your child's maths from a school exercise book. If there are pages of sums ticked in red that means that your child is ready to move onto the next stage. If much of his school maths work is marked wrong, that may only mean that he has difficulty in writing maths down. You can help your child by putting the maths that he is doing into practical contexts in order to help him to understand it better.

Helping with fractions

Children get very stressed by fractions! When I am teaching classes of nine- to 11-year-olds it is fractions that seem to confuse everybody, including the very bright children. I think that colouring $\frac{1}{2}$ of a square, or $\frac{1}{8}$ of a hexagon is perhaps not very meaningful to them. They need to have a lot of experience in halving apples, slicing up pizzas and cakes and sharing out nuts and chocolate to be able to put the fractions of their school text book into some context that they can understand. Those maths work books that you can buy in the supermarket or stationers can be a help, but the best ones will be those with practical activities, not just rows of sums.

You will see from your child's maths book at school that there is much less emphasis nowadays on sums like this.

$$4\tfrac{5}{7} + 2\tfrac{3}{11} =$$

You can probably remember doing pages of those at school! We don't often have to use sevenths or elevenths in our daily life now so where fractions are used, they tend to be the more usual kinds, such as halves, quarters, and thirds and others are added for fun and to help understanding. Increasingly we use decimals in our daily lives now, so they get more emphasis than fractions.

Activities to help with fractions

Here are some activities that I have used over the years with confused children (and adults!).

The zero to 1 number line

This is a very simple idea but a revelation to many children. All fractions and decimals are somewhere alone the line from zero to 1 (but of course $1\frac{1}{2}$ is beyond 1). So putting the fractions that we are talking about along this line helps to build up some mental image for the child of what $\frac{1}{3}$ or $\frac{1}{100}$ actually means.

Making a line for the kitchen or bedroom wall can be very helpful. You need a long bit of paper – a bit of old wallpaper will do, and a thick felt tip. Draw the line from zero to 1 and put on it some of the fractions that your child knows about. Ask them where $\frac{1}{2}$ comes. Where would they put $\frac{1}{3}$? If they are confused about this, do the paper folding activity on page 151 before you go further.

The number line can be added to as a new fraction is encountered, e.g. if the 1 is £1, where would 50p go? Where would 0.2 belong?

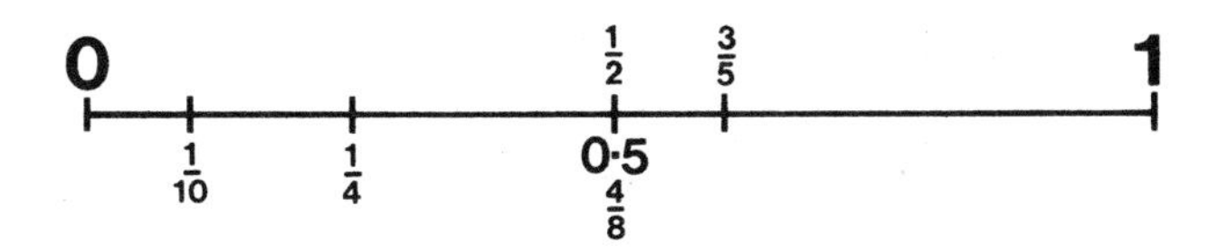

Fractions are divisions

Make sure that your child sees that fractions are only another way of writing divisions. They are not some very different kind of maths that he is never going to understand – which is what many confused eleven-year-olds tell me.

So if you thought of 3 divided by 5, you could write it like this

$5\overline{)3}$ or you can write it as $3 \div 5$ or as $\frac{3}{5}$.

So three-fifths is no more daunting than three slices of a pizza that has been sliced into 5 equal bits.

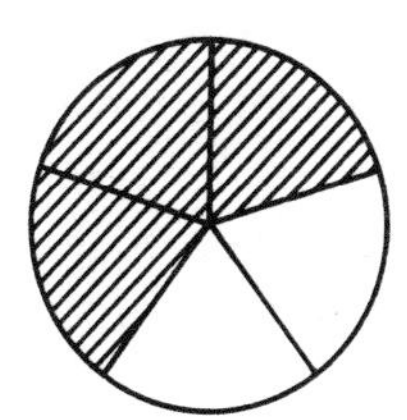

Paper folding

This is a very easy activity and can help children feel relaxed about fractions. They understand how to fold a piece of paper in half but they may not yet have related that folding to that mysterious symbol of $\frac{1}{2}$ that appeared at school.

Fold a piece of paper into several strips.

1 whole							
half				1/2			
quarter		1/4		1/4		1/4	
1/8							

Name the top one '1 whole' and remind your child that all fractions and decimals are a bit of a whole one. The whole one might be a pizza, or a sausage, or a class of children, or a bit of paper, or anything that can be split up into bits.

Now take a felt tip and split the next folded section in half. You could write 'half' in one part and $\frac{1}{2}$ in the other. Split the next section into quarters – that is, half of a half. Then split the next section into eighths and so on to the bottom of the paper.

You can now see from your paper that two quarters are the same as one half and two sixteenths are the same as one eighth. Put some of these fractions onto the zero to 1 number line. You can show your child how a thirty-secondth, (or thirty-twooth as I call it) is much smaller (and therefore nearer zero) than one quarter.

You can put your bit of paper up on the kitchen wall and another day make a different one. You can make the second line divide into thirds, or fifths, or sevenths.

The shade-in game

I learnt this game when I was studying on an Open University maths course for teachers and I have taught it to parents, teachers, students

1 whole					
third		1/3		1/3	
1/6	1/6				

and children over several years. It is successful because it is fun to play and because it teaches that difficult area of equivalent fractions (that two quarters is the same as one half).

You need:

- pieces of A4 paper;
- felt tips;
- a big cube or some slips of paper in a pile.

Fold a piece of A4 paper for each player into sixteen pieces. Mark the cube (or the slips of paper) with some fractions such as:

- one sixteenth
- two eighths
- one quarter
- two sixteenths.

I usually write it out in words like this at first as children are sometimes confused by the symbols. We need to teach the concept of the fraction first: the recognition of the symbol is a later skill.

Each player has a piece of paper and takes it in turns to throw the cube or to turn over a piece of paper from the pile. If she turns over, say, one sixteenth, she has to explain to her partner how much of her paper is one sixteenth, so she might say,

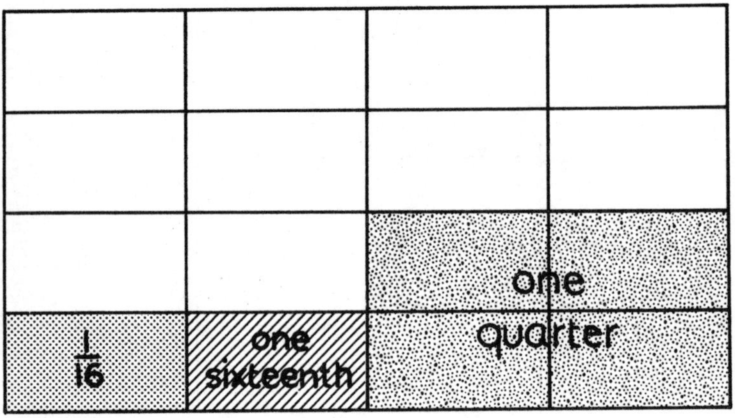

$\frac{1}{16}$ one sixteenth one quarter

'this little section here is one sixteenth because there are sixteen of them on the whole bit of paper, and so one bit is one sixteenth'. If her partner agrees, she is allowed to shade in that one sixteenth.

The game goes on, with players taking turns to explain and shade in a section. The winner is the one who shades in all of his paper. If you throw a quarter and don't have a quarter left, you miss a go.

This game is good because it helps children to see the equivalence of fractions (that a $\frac{1}{2}$ is the same as $\frac{2}{4}$) and to appreciate some of the language of fractions. Understanding the concept of fractions can only come if the child can understand the various words that are used and what $\frac{1}{16}$ really means. You can obviously play the game with different fractions by folding the original piece of paper into a different number of sections instead of 16. The game can be played with a younger brother or sister from about the age of seven.

Other paper-folding activities

Try folding a piece of paper as many times as you possibly can. What would be the name of one of the sections?
You might be able to make one section into a sixty-fourth. If you have a bigger piece of paper, can you fold it more times?

Origami is a really great mathematical paper-folding activity for

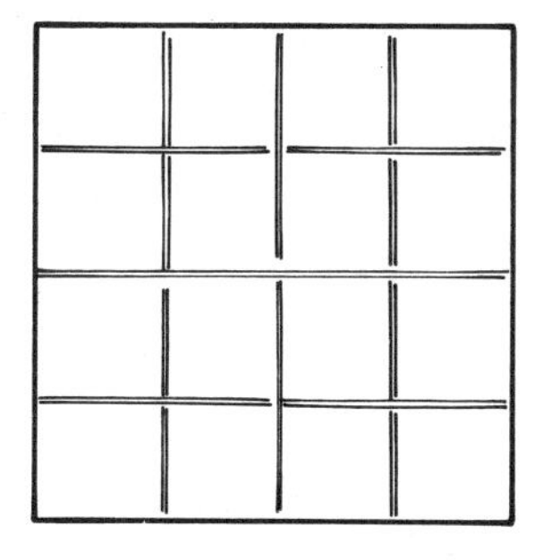

Only two folds by two folds are needed to fold this piece of paper into sixteenths

this age group. Buy a good origami book, or borrow one, and some coloured paper. (You can get special origami paper, but anything will do to get started.) Toy shops now have pads ready marked to make models easily. These are great as starters. If you work with your child that will get her over the stage of working out what the various folds mean. Origami figures make great home-made birthday cards.

Making snowflakes for the window at Christmas

It's easy to make eight-pointed snowflakes, but real snowflakes have six points. You need to cut a circle of paper, fold it in half, then fold it very carefully into thirds. Cut it out and stick it up on the window with a bit of Blutack or sellotape.

Sausages

This is a good activity to teach fractions, but can also be developed into a game for division.

There are eight sausages in a packet and five people in the family who all want an exactly equal share of them. Make the sausages out of paper – get the child to draw and cut them – and share them out into five equal groups.

It's quite complicated!

Help your child to name the 'bits', e.g. when something is divided into ten bits, each bit is a tenth, if it is divided into 22 bits it is a twenty-second, but if you call it a twenty-twooth your child is more likely to remember the activity.

You can play this as a game, making it as complex or as easy as you want it to be. You could have 12 sausages and three people, or 19 sausages and 12 people. As above, the idea is to give each person an exactly equal share.

Helping with tables

There is advice for helping with tables in chapter 8, but if your child is still having problems at this age you might want to set aside a couple of 15 minute sessions each week when you work together on tables. All the activities in chapter 8 are still appropriate at this stage. If you have tried a number of activities and want something different, Tarquin (see page 190) do some table cubes for under £5 which are excellent for learning tables facts; they also produce a tables colouring book for under £3. Both of these will help if your child's need is to be able to memorise the facts, but do check that your child is understanding what multiplication is all about, and help him to see the connections between the tables. (So if they know 6 lots of 4 is 24, they know that 3 lots of 8 is also 24.)

I have found multiplication versions of 'four in a row' games (see page 126) very successful with this age group.

Helping with decimals

Decimals are about dealing with fractions that are tenths, hundredths and thousands.

For children who have had a great deal of experience with a calculator, decimals have been part of their maths for some years, but for those who have not had this experience, this can be another problem area.

Make sure your child can:

- multiply and divide by 10 and 100;
- appreciate that if £1 is the whole one, the pence are the decimals, so £1.50 is one and a half pounds, or 1.5 pounds;
- use an abacus;
- read a decimal number off a calculator, so 1.75 is read as 'one point seven five';
- use a calculator correctly, so 13 divided by 2 gives the answer 6.5, not sixty-five. Does she understand that 6.5 is six and a half? So does she see that 13 biscuits shared between two is six and a half each?

Decimals and money

If you want to help with decimals and you have an old bead frame abacus in the toy box, this is the moment to dig it out.

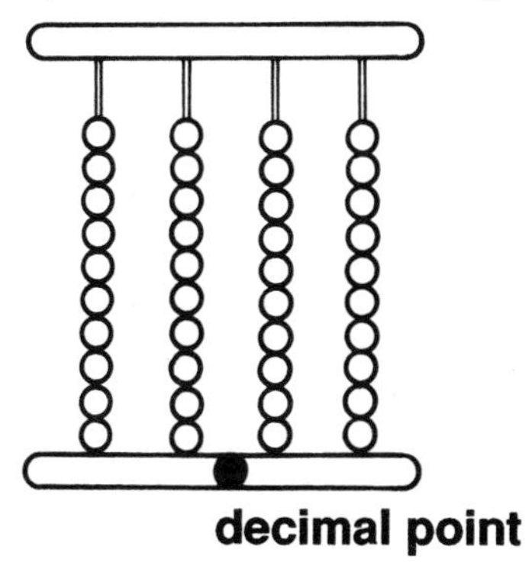

Mark a place on the abacus for the decimal point with a coloured dot. So everything to the left of the decimal point will be whole pounds, and everything to the right will be pence.

Now think of some prices that your child can put on the abacus. For example, a packet of their favourite breakfast cereal costs £2.17.

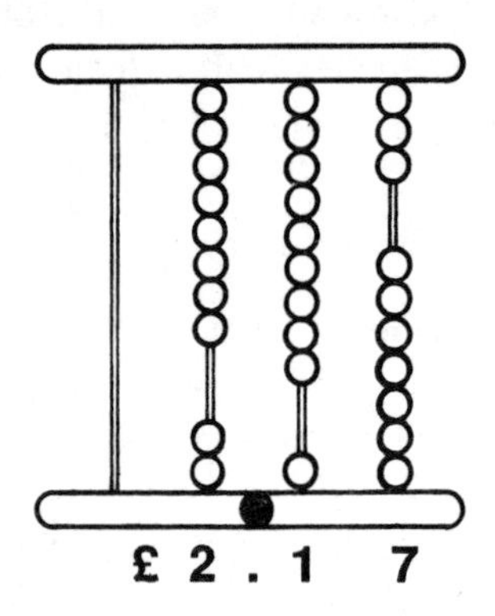

That means two whole ones, and 17 hundredths. Setting it out in real money might help it to click that it is 17 hundredths because each penny is a hundredth part of a whole pound.

If your child is still very confused after you have done this a few times, check that he has understood place value (page 89).

- Check that he understands that 50p is 0.50 (or 0.5 or a half) of a pound.

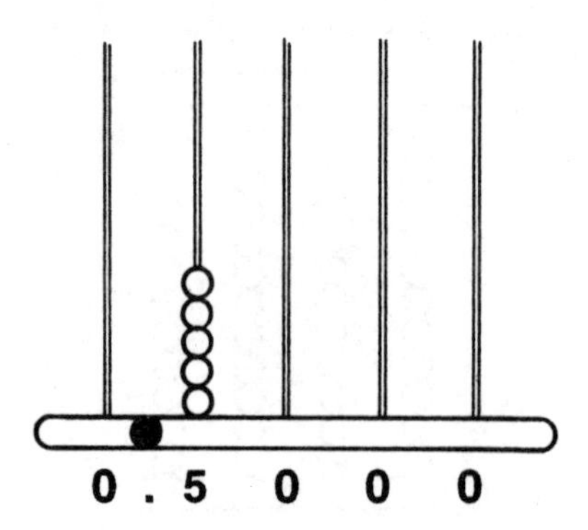

- Help him to learn the important decimals 0.25 (a quarter), 0.5, and 0.75 (three quarters).
- Check that he sees that 0.5 is the same amount of money as 0.500. Put this number on an abacus. If you don't have an abacus, getting him to do a drawing like this one is helpful.
- Check that he can multiply 1.5 by 10, and 6.7 by 10 and any other number by 10. Let him do it with a calculator first to see the pattern.

Games to help with decimals

Target

You need two players and a calculator.

You agree on a target, say, 100. You can only use the keys **0** to **9**, the ×, the =, and the decimal point.

The first player puts a number into the calculator, say 39.

The second player must now try to turn that into 100 by multiplying:

$$39 \times 3 = 117$$

Now the first player must try to make 100 from 117 by multiplying:

$$117 \times 0.8 = 93.6$$

Then the second player takes 93.6 and tries to make 100 by multiplying, and so on until you get as close as you can to 100.

Three in a row

For this game you need a number line drawn from zero to 1 on squared paper (or graph paper if you have it), a calculator and red and blue felt tip pens.

First draw the number line and then take turns to choose two numbers from 1 to 15 to make a fraction (say, $\frac{4}{15}$, or 4 divided by 15). Use the calculator to convert the fraction into a decimal (put in $4 \div 15 = 0.26666$) and mark this decimal on the number line in the right place and with your colour.

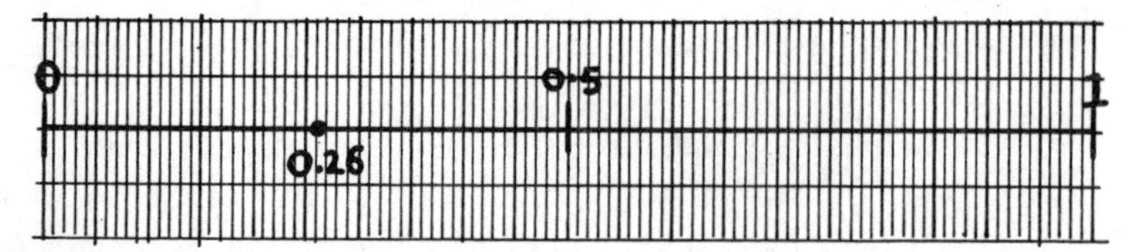

When you have both had a go, you both have another turn, but the aim of the game is to get three of your coloured blobs in a row, without any of your opponent's marks in between your marks. It's not easy!

You can make the game even harder if you agree that if you make a mistake and end up with a decimal fraction that is greater than one (so it won't fit on the zero to 1 number line) you miss a go.

Helping your child with 'media maths'

Maths is used in newspapers, on the news and in advertisements. Helping our children to understand this maths will help them in their adult life.

- Pie charts and graphs make good discussion points. Does the graph distort the truth by using a misleading scale?
- What does it mean when we are told on some advert that nine out of ten cats prefer some cat food? How was that statistic arrived at? Who tested it? Did they ask the cats?
- How are things like opinion polls actually carried out? Do the questions people are asked lead to biased answers?

These are the sorts of ideas you can discuss with your child. It is important that she is not taken in by the media and the so-called 'facts' that we are confronted with each day.

Understanding percentages

Amongst the statistics that we see in newspapers are percentages. These are a bit like fractions and a bit like decimals – because percentage means 'out of a hundred'. So:

50% means 50 parts out of 100.
or

$\frac{50}{100}$ or a half.

So:

$\frac{1}{2}$ of 100 = 50%.
$\frac{1}{3}$ of 100 = about 33%.

That's quite easy – well – it might give you the panics and remind you of school. It's doing that to me as I write this! The real problem comes when we are dealing with something like the percentage of people that have a cat if 35 out of 170 people asked said they owned one (I made these numbers up).

Now I have to think back to the way that I got 50% (1 part out of 2, times 100). So the percentage of people with cats is 35 out of 170, times 100 – because percentage always means 'how many out of a hundred?'.

So that is

$\frac{35}{170} \times 100$.

At school I would have panicked about this, but now I ask myself, 'what is a reasonable answer to this?'. I'm going to think about that by first working out how many 35s there are in 170. It must be something like 5, because there are almost 3 lots of 35 in 100, and then another 2 lots of 35 to get up to 170. So my answer will be about 5 parts out of 100, or somewhere around 20%.

At school your child would then work that out on a calculator – so would I. I make the answer 20.588235%. I would round that up to 20.6% so my estimate was not too far out. It isn't a 'bad' thing to work it out on the calculator. The hard work is knowing what to do with what number!

Almost all this chapter has been about number because that is the main area in which children seem to need help at this stage. In the next chapter we will explore some of the other areas of maths and it is these activities that can be influential in helping your child to feel that he can do maths and that he enjoys it.

CHAPTER ELEVEN

Things to do with your nine- to 11-year-old

Enormous numbers

By this age children can deal with large numbers and are usually fascinated by them. There are several ways that you can help them to have fun with these numbers.

- A calculator is an essential. Borrow a book of calculator activities from the children's library. Many of them are very good.
- Get a book about stars and planets. These are full of enormous numbers for your child to explore.
- Get your child to think of some problem that he could ask friends. What about how many minutes they have been alive? Is it nearest to 1 million (1,000,000), a billion (12 zeros), or a trillion (18 zeros)?

Patterns from nature

One of my own personal passions is the pattern and design that we see around us in the natural world. There are patterns in wild flowers, in birds, trees, insects, the constellations of stars at night and in the phases of the moon and the tides.

Looking for spirals

Spiral shapes are some of the most interesting around. They are there in:

- snail shells;
- the pebbles on a beach in Dorset, where almost every one is a spiral ammonite;
- fir cones;
- sunflower seed heads (these are easy to grow if you get them planted in good time in the spring. Provide a cane for support and water them in dry weather);
- the patterns on the skin of a pineapple.

The hexagons on the skin of a pineapple are all arranged in spirals. These spirals go at different angles around the pineapples and each hexagon is on at least three different spirals. It needs a felt tip pen to identify them. An extraordinary thing about these spirals is that if you looked at a hundred pineapples, almost every one would have either eight, 13, or 21 spirals. These numbers are some of the most exciting numbers in the universe! That's no exaggeration. These numbers are called the Fibonacci numbers and they crop up all over the place in the natural world.

Fibonacci numbers

If you are interested in this, it is worth looking up Fibonacci in a good encyclopaedia. It might also come under:

- the golden ratio
- the golden rectangle
- Pascal's triangle.

This would be a good project to help a child to learn to use the encyclopaedias in the library.

Pascal's triangle

Here is a good starting point for a bored child. Can you continue this number pattern?

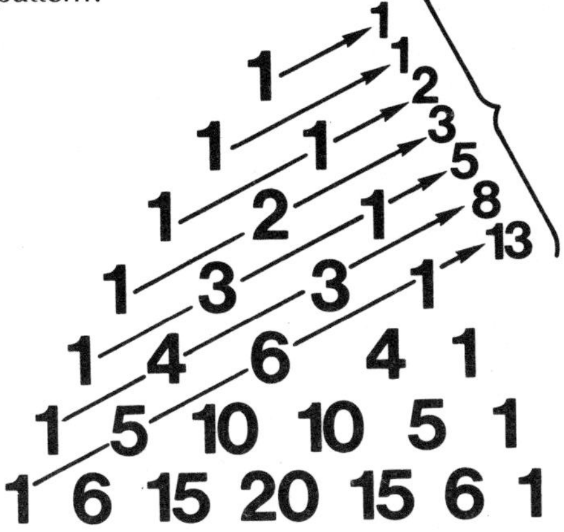

Can you see how the numbers are generated? The pattern begins with 1 and each number is the sum of the two numbers above it. It goes on and on and on – have a large piece of paper ready. Your child will by now be able to talk about infinity – the idea that numbers go on for ever.

The numbers on the diagonals in this triangle are the Fibonacci numbers – those ones that pop up all over the place in nature.

Other patterns in nature

We look at the petals on flowers to see which flower we are looking at, but petals can tell more than just the species of plant. It is a great help to a child to be able to put flowers into families. All the cabbage family, for example, have four petals; these are called the Cruciform flowers – shaped like a cross. Common examples are rocket and shepherd's purse and, if you are on the beach, sea rocket. (Not all flowers with four petals are in the cabbage family, of course!)

Many flowers have five petals. Examples of this are buttercups and celandines. It is important not to pick wild flowers, but a maths trail in the countryside or a park could include:

- Find and draw a flower with five petals.
- Find and draw a flower in which some petals are bigger than others (vetch, broom, clover, etc.).

Apparently more flowers throughout the world have five petals than any other number and a maths trail could include looking for fives in nature:

- a sycamore leaf has five points;

and on the beach you might find

- a starfish;
- a fossilised sea urchin if you are near chalk cliffs.

Five (a Fibonacci number) and five-sided (pentagons) are common in nature.

Symmetry from things around us

Many things around us are symmetrical (some flowers and leaves for example) and children can explore this with a mirror, or they can look for rotational symmetry (page 170). One way a child can explore symmetry is for you to give her half of a picture from a magazine and suggest that she draws in the other half. It might be half a face, or half of a space ship, or half of a building.

Maths in art

If you or your child enjoy art, there is a great deal that you can do to encourage him to explore maths through it. Many of the Tarquin books (such as *Make Shapes*, *Symmetry Patterns*, and *Tilings and Patterns*) appeal to an artistic child. Contact Tarquin Publications at Stradbroke, Diss, Norfolk, IP21 5JP.

There is a good book, too, called *Mathematics through art and design 6–13* by A Woodman and E Albany (Unwin Hyman).

You might already have a Spirograph set; the patterns it makes are rich in maths. Work with your child and see if you can both predict the number of 'points' a shape will have before you draw it.

Maths from a geometry set

If you haven't already done so, it is a good idea to buy your child a geometry set (or maths set). There is a section on this in chapter 9, and by age nine your child can be quite proficient at designing things for herself and at using the compasses, set square and protractor. She will need to use these things in secondary school and making patterns with them now is a good way to get used to them.

You might want to collect your child's work in a folder or scrap book, or give one wall in their room over to putting up designs and posters. You can stick cork tiles onto a wall with rubber-based glue to make a cheap and easy notice board, or just resign yourself to redecorating the wall when he is 18 and doesn't want it still covered in Blutack.

Children are tremendously creative if we provide them with the right equipment and encourage them to switch off the television. Here are some designs that one group of nine- to 11-year-olds did.

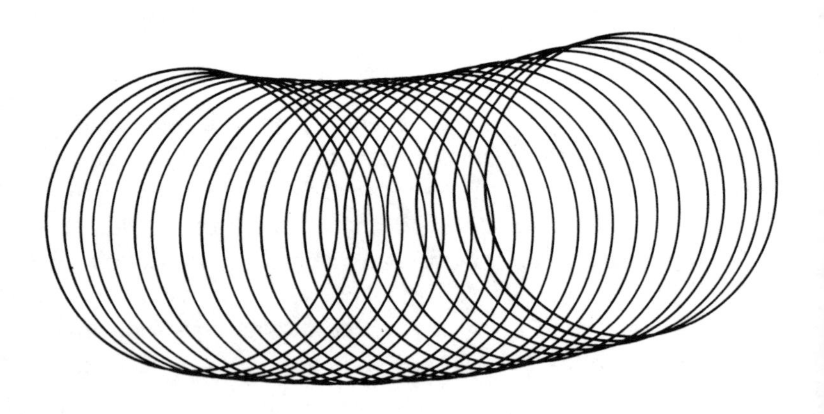

Tariq (10) drew circles like this and said, 'It's like going down into a black hole'. Later he wrote a story about a space mission to rescue someone from a black hole, and after that he drew a spiral on the computer with Logo

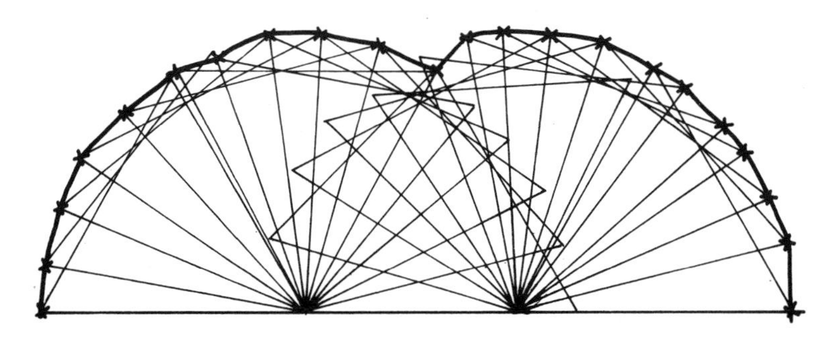

Alice (9) made a triangle out of card and then 'walked' the triangle along a line. She marked one corner of the triangle with a cross, then marked the path of the cross. I asked her if she did this with a hexagon, what shape would the path of the cross be? She made a prediction and tested it out

Yasmeen (10) made this pattern (below, right) with concentric circles and parallel lines drawn with a ruler

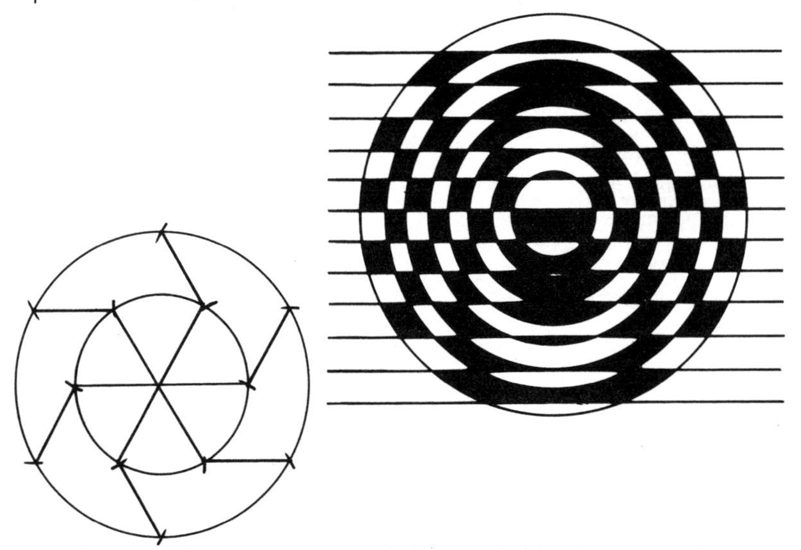

The basic circle pattern (above, left) is made like the one in chapter 9, but instead of drawing the arc in the circle, marks are made on the edge of the circle and then these marks are joined with straight

lines. I call it a 'basic circle pattern' because once children can do this, they can do hundreds of different patterns.

Many of these patterns have rotational symmetry. That means that if you put a pin through the middle of this picture, you could rotate the picture and it would still look the same. As you rotate it, there are six positions that this circle can be in to look the same before it gets back where it started from.

You can see that this kind of symmetry from the basic circle pattern is different from the sort that you do with a mirror. Remind your child that if they want the circle pattern to have rotational symmetry they will need to think very carefully which colours they use and where they put those colours.

If your child needs more help with rotational symmetry, get out a pack of playing cards. If you hold the king or queen one way up, then rotate it for half a turn it still looks the same. You get a similar result with the letter 'S' cut out of card.

Other fascinating circle patterns are 'mystic roses'. To draw these, you start with a circle and the mark points around the edge, perhaps every 30 degrees. Then all the points can be joined to all the other points which leaves you with a beautiful complexity of lines and intersections. These are good to make with thread too, using wood and small nails around the edge of the circle. These pictures make unusual presents.

Tessellation patterns

If you have been to the Science Museum in London, you might have seen in the shop some of the designs of the artist M C Escher. They get onto ties, tea towels and sweat shirts and they are really fascinating. Some of the designs are illusions and impossible figures, and others are remarkable tessellations (shapes that fit together).

Children are fascinated by tessellations and if you have been on a maths trail, you will have found them all over the place – the tiles in the bathroom, patterns on a jumper, paving stones in the street, etc.

Your child will have had experience of tessellating plastic shapes at school and will be able to tell you that hexagons tessellate (as they do in a honeycomb), but circles don't fit together (so we don't put circular tiles on the wall). You can give your child further experience with tessellating shapes by giving her card, sellotape and scissors, and encouraging her to try this activity.

1. Starting with a shape that tessellates (squares and rectangles are the easiest – hexagons and triangles are a bit mind-blowing and only for the bold), cut the shape carefully out of the card.
2. Mark off a section on one or more edges and cut it off.
3. Then *without rotating any of the bits of card*, move the cut-off section(s) to the opposite edge and stick it down with sellotape.
4. Now, if you try to tessellate your shape, you will find that the bits you cut off will fit into the 'holes' you made in the shape.

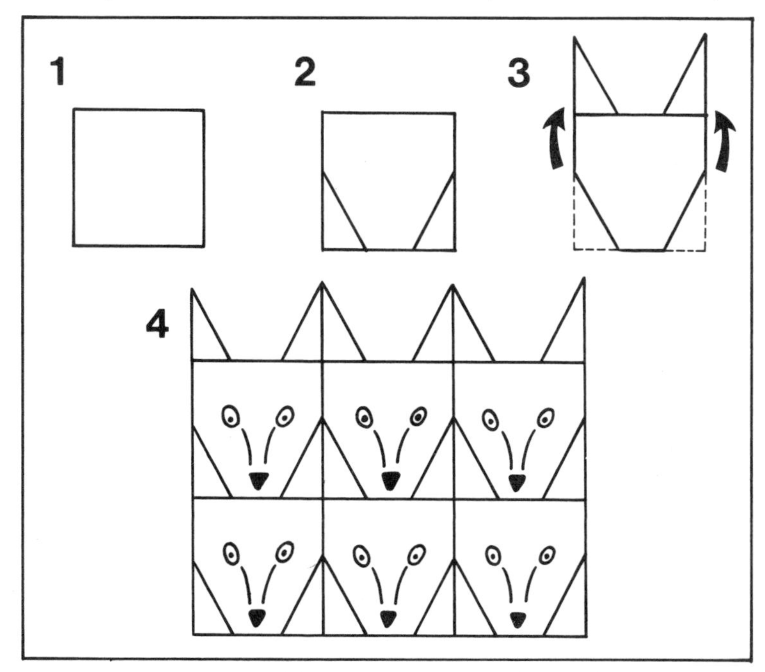

Simon (10) made this fox. Others made cats, penguins, sharks and monsters

Tarquin (see page 190) sell Escher posters of tessellations and also 'kaleidocycles' – 3-D shapes that you can make at home that will rotate and surprise you.

Overlapping edges

If you cut a shape out of card, you will find that you can make exciting patterns by overlapping it and drawing round it in different positions. Or you can cut several of the shapes out of tissue paper and stick them down, overlapping them. If you use more than one colour of tissue paper you will make a picture good enough to keep or to use on a card.

Learning to solve problems

By this age, children can do quite sophisticated things on their own.

- Design a treasure hunt for a friend. It adds interest to this if you let them dye a piece of paper with cold tea to make it look old like a pirate map.
- Run a fete in the garden or the street to raise money for charity. Working out how to make games at which they will make money takes a lot of thought. For example, if a ping-pong ball lands in one of the jam-jars you win 10p. You have to test how likely it is that people will get the balls into the jars otherwise you will lose money.
- Make a toy out of wood for a younger child, or a box with a lid for grandad's birthday.
- Make a time line for their bedroom wall that is to scale and shows important historical events. It's easiest for them if you provide squared paper to do this, and if you don't want Blutack on all the walls suggest that they do the time line as a spiral.

Games, toys and books

By this stage the games and toys are getting fascinating. Don't try to resist the urge to play with them yourself! Join in and have fun!

My own children had a great deal of fun with the Tarquin publications (see page 190) and the other kinds of books and games that are available from places such as the Science Museum in London or the bigger toy shops. Larger toy shops have wooden puzzles which are often sold as 'executive toys', such as the Tower of Hanoi, or cubes that interlock and seem impossible to fit back together again. Stationers have books of puzzles. All of these things can help your child to see that maths can be interesting and challenging.

There is so much that children of this age can do, from joining a chess club to making a sundial. In these years just before secondary school we can really help them to feel good about their mathematical abilities.

There are many mathematical books that are fun. Johnny Ball (*Johnny Ball's think box*, Puffin) is great for fun activities for the more able child. Usborne do several books that are well illustrated and easy to follow, such as *The Usborne pocket calculator book* by M Burns.

If your child needs help with some basic number work you might find that the Usborne *Multiplying and dividing* books help. One is a write-in work book and the other is just an information book.

One of my favourites is *The I hate mathematics book* (M Burns, Cambridge University Press). It really is fun but you might need to work with a child on it. It has a good chapter on maths for a child to do if he has flu. If you are a bit nervous about maths yourself you could get hooked on this book! Other good books are:

- *Cards on the table: 20 mathematical card games* by Fran Mosley, BEAM. A good book for maths at home.
- *Starting from calculators* from BEAM. Other 'Starting from . . .' books are going to be published.
- *The amazing mathematical amusement arcade* by Brian Bolt, Cambridge University Press. Brian Bolt has written several books of maths puzzles and activities, and they all have the answers in the back.
- *How puzzling* by C Snape and H Scott (Cambridge University Press).
- *How amazing* by C Snape and H Scott (Cambridge University Press).
- *A way with maths* by N Langdon and C Snape (Cambridge University Press).

CHAPTER TWELVE

The transition to secondary school at 11, 12 or 13

The age of the move to secondary school varies enormously but for all children (and their parents!) it is a huge step. There are some changes that you can help your child with, but no doubt others they must face on their own as a part of the growing up process.

What to expect from secondary school maths

Your child enters the National Curriculum Key Stage 3 when she is 11, which means that she will continue to be taught from that curriculum until she is 14, when she will start her GCSE work. Entering Key Stage 3 will inevitably mean harder work, but it should mean a certain amount of continuity. Records of the level of your child's attainment will be passed on when she changes school and, theoretically anyway, she should be able to continue the maths curriculum from where she left off at primary school. Many schools seem to have the policy of going over all the maths that they think the child should have done at primary or middle school. This might be good for your child if she needs a recap. Equally, you may find you have a year of her being so bored that she learns to hate maths.

It might be that the passing on of records improves because of the National Curriculum. When your child leaves his primary school it might mean that, having sat a national test, the teachers will know more of what your child can do. Then again, testing is notoriously poor at telling us what a child can actually do, so the situation may not improve as much as some people hope. In the end, every teacher likes to make her own judgement about a child and testing does little to tell a good teacher anything that she couldn't work out for herself given a few lessons with your child.

A problem that might arise

One problem that you may well find with the National Curriculum levels is that if your child is, say, at level 4 at age 11, she may seem to progress very little and still be given a level 4 at the testing at 14. It seems that (on the specimen papers we have at the moment) a level 4 question in an age 11 test is considerably easier than in a level 4 test at 14. So it might look as if your child has learnt nothing in two or three years! (This type of problem probably wouldn't arise in the Scottish testing system.)

There is more about the National Curriculum and tests in the appendix, but don't expect testing necessarily to improve a school or what your child does. My own rule of thumb is that tests usually tell you what that particular child did on that particular day with that particular question – and not a lot else.

Big changes

The biggest change for your child is that he will cease to have the same teacher for everything and will have a teacher for each subject. Unless he experienced this at middle school, it is quite stressful to have to cope without the familiar primary teacher who is there all the time. Most children handle this change well, but it does mean that the teacher probably teaches more than 200 different pupils in a week. The personal touch of the primary teacher is almost impossible at secondary school, and it is now that we need to be even more vigilant about our child's progress and difficulties. At the same time, though, our children are more independent and less willing to discuss school work, so it's harder to keep in touch.

A child may not be willing to admit to needing help within a class where maybe there are only a few familiar friends, a new teacher and the need to keep up the 'really hip and cool' image of the peer group! If she admits to us that she is lost, she may well plead with us to say nothing for reasons that are obvious to our child, and that we have forgotten we also felt at the same age.

Maths becomes more book based

Inevitably the maths becomes more linked to text books at this stage. Many schools are enlightened enough, though, to keep using maths apparatus. Many of the adults I work with say they could not manage without cubes, or something to represent what they are doing. How much more true that must be for the 11- to 16-year-old.

Maths ability groups

Putting children into ability groups is a common feature of secondary schools – inevitably so, as by now the gap between the most able and the least able is enormous. There is a positive and negative side to ability groups.

The positive side of groups

- The children can work at their own level and pace.
- Often lower ability groups will be smaller so the teacher can give children more attention.
- Bright children can be challenged and can do maths that is exciting for them.
- Teachers who take lower ability groups are sometimes

especially gifted at helping children to succeed and to feel better about themselves.

The negative side of ability groups

- Children can feel a stigma at being in the 'bottom group'.
- Some children feel labelled as 'bottom group', which can mean that they then behave as 'bottom group'.
- Teachers can also label children and respond to the 'bottom group' with so little expectation that, of course, little is achieved.
- Some schools let groups be too static and if your child is put in the third group, that may be where he stays for the rest of his school life. Children develop in very individual ways and they can change from being hopeless at maths to getting a very respectable grade at GCSE. Ask at school for the policy on changing groups. A good school will have a clear and flexible policy so that a child can change group as circumstances change.
- Sometimes children can be put in the 'wrong' group for them. A child may be terrified of the pressure of 'keeping up' if put in the top group.

Making space for homework

Where we can help is by making a space for homework. I mean that in the physical sense of a desk with a lamp and a comfortable chair and maybe bookshelves, but I also mean it in the sense of making space in your child's day.

This may mean changing the habit of television watching and making rules that will help your child to settle into the routine of homework. (If you want to read more about settling your child into the secondary school routine see chapter 11 in another book in this series, *Help your child through school*.)

Keeping links with your school

It's really hard for many parents to keep up the links with school at this stage. For most of us it is a time when we can get work outside the home and there just isn't the time to go to the school, and the schools often don't encourage parents to come to them. Our children don't want us to go to school with them and so it is all too easy to find that the only time we see the teachers is for parents' evenings when we may discover that there has been a problem we knew nothing about.

Children's lack of confidence

In my recent research into people's lack of confidence with maths, the move from primary to secondary school is seen by many adults as the time when they really began to fear maths. Many people cope with maths at primary school, but get lost with the xs and ys of algebra and equations, and the other maths that seems to many of us to be complete gobbledygook!

If your child begins to show signs of anxiety about maths, first discuss it with her and see if there is anything that you can do to help. It would probably help to discuss the situation with the teacher, but children are wary of our intervention and may request that we don't contact the teacher. There's no easy answer to that!

The problem with maths anxiety is much worse in girls than in boys. My secondary colleagues comment on the ways that girls often put themselves down and insist that they cannot do their maths, when they can really, but they don't believe in themselves. It is this lack of confidence that starts them on the move down from higher to lower groups, and poor exam results at 16. I'm sure my secondary colleagues are right when they say that they know they have to put considerable effort into encouraging girls. Of course, they encourage boys too.

Maths and self-image

One of the things that is most striking about the research I have done over the past few years is the way that people's self-image, or self-esteem is bound up with their achievement in school and in life generally. Sometimes people perceive themselves as 'no good' because they cannot read, or cannot get GCSE maths, or don't understand something.

It is vital to our child's future well-being and personal growth and maturity that he sees that it is who they are that matters, and not so much what they achieve. We care for him and love him no matter what happens. He does not become a *better person* if he does well at school. It is all too easy for home and school to give out the message that all that matters is to achieve. The child going to university is somehow thought to be 'better' than the child who wants to go straight into a job. She is maybe going to have a 'better' education, but that does not make her more valuable or a more loveable person!

It seems to me that it is crucial to give our children the message that it is who we are that matters, not so much what we achieve.

'How can I help my child with maths when I don't understand it myself?'

It is a real shock to the system when a child brings home his maths homework and we haven't a clue what the question means – let alone how to find an answer! This is a real problem for parents and children and there is no easy solution. What I know has worked well for some of my friends is to find someone who will help with homework if they are really stuck. A friendly baby-sitter, or an aunt or neighbour can often give the help that is needed. Of course, really it is the school that should become involved, but children are understandably quite resistant to that.

It is helpful to discover what the school policy is on homework. Some schools are only too happy if you get the child started on a difficult problem, or do the work with them, but some schools like you to put a note at the end to say what you have done.

In some schools all the assessed course work is done in school in front of the teacher. In other schools assessed work can be done at home. These are the parts of the GCSE syllabus that your child does in the year before the exam and which are marked by the teacher and count towards the final examination grade.

I found that it was at this stage, much more than in primary school, that it was helpful to buy the text books my child was working from. I didn't understand all the maths, but it was helpful to have it all in front of me and for my child to have access to the book. (School copies couldn't be brought home because there were not enough copies to go around.)

You can often find the relevant text books in ordinary public libraries, or you can ask them to get them for you. If you can afford it, you can buy copies of the text books in high street shops – if they haven't got them in stock they will order them for you. I found this a slow process though and discovered that, provided I had the name of the publisher and the title and number or level of the book I wanted, I could ring up the 'customer services' desk of the publishing firm and they would send me the book immediately and invoice me. This often saved the day.

As well as the school text book, there are some other books that are helpful at this level. Anything in the resources section on page 174 would be useful. In addition you could try:

- *Sources of mathematical discovery* by Lorraine Mottershead (Basil Blackwell);
- *Investigations in mathematics* by Lorraine Mottershead (Basil Blackwell).

The two Lorraine Mottershead books will give you and your child greater insight into the type of maths now taught for GCSE. They are widely used to supplement school maths schemes. Brian Bolt's books are great fun – and have the answers at the back. They are particularly useful if your child likes puzzles but is less than enthusiastic about maths.

There are several Tarquin books that could help a child to feel that maybe maths is quite fun. There is a series of *Mathematical curiosities* – models to make that are surprising and fascinating. The *Make shapes* books make impressive mathematical models, including a large 3-D star that makes a good Christmas decoration. There are so many Tarquin books suitable for this age group that I can't list them all, but your child might enjoy choosing something from the catalogue. Contact Tarquin Publications at Stradbroke, Diss, Norfolk, IP21 5JP.

Parents are the child's best teacher

We might feel that secondary school teachers are the experts and that we should leave it all up to them, but at all stages of a child's life it is really the parent who can be the child's best teacher and the person who influences them the most. Don't be put off by bossy and 'I know it all' teachers! It is *you* who knows how your child learns. All children seem to learn in slightly different ways and you might need to find a tactful way to tell a teacher that – if you have the confidence!

I found secondary school parents' evenings terrible occasions and some of the teachers were very intimidating. Some teachers like to give out the image of power! It immediately reminds us of the worst teachers we had as children and all the things that we want to say to them about our child disappear. If something is very important, put it in a letter.

Coping with absences

My secondary school colleagues comment on how many children still go on holidays with parents through the school year and the considerable impact that this has on the child's work. When we take our holidays is not always something that we decide, but though it makes very little difference to a child at primary school, it seems that a fortnight of missed lessons once the child is doing more advanced work can be a bit of a disaster.

Absence through illness is obviously unavoidable, but it is a good idea to get one of your child's friends to bring notes on missed lessons for your child to copy up, and to bring any messages or homework assignments. Maths is one of those subjects where if you miss something vital along the line, the rest of the learning for the next month can get all confused. Talk to your child's teacher if you are worried, or ring up the school secretary and ask for work to be sent home. A good teacher will always help your child to catch up.

When do I need to get extra help?

The secondary stage is sometimes the time when we need to get outside help. I think if your child is falling behind, or is anxious at secondary school, it is money well spent to find a good tutor. The school does not have to know what you are doing and it seems that many young people don't want the school to know – and that applies to any subject – not just maths.

As parents, we are in many ways the child's best teacher, but sometimes it is hard to help our own children – especially if we don't understand it either! Sometimes our children resist our efforts to help and in order to keep our good relationship with the child, we need outside help.

Finding a good tutor can be a long and complex business. Asking around was the most successful way that I found, and that way I got someone who came with a good recommendation. People who have put their children in for exams at 11 or 13 are often a good source of knowledge about private tutors, or you can ring up a nearby prep. school. Sometimes local libraries keep lists of tutors and local papers often advertise private tutors and agencies.

The tutor needs to be someone kind and friendly, who will relate to your child as an individual and who can meet your child's need to be encouraged as well as be taught the facts. I'm not sure that it matters if the tutor's methods are very different from the school's. Some people feel that this may confuse the child, but I think that every teacher teachers in a slightly different way. Getting a different slant on what puzzles the child is what you are after anyway.

My child is good at maths – what careers are there other than accountancy?

Alongside all our everyday uses of maths, there is more specialist maths that children start to study later on in school. There are a number of careers that are open to children who are good at maths. As well as the obvious ones like banking, maths is very important in all of the sciences. So careers in physics, medical sciences, weather forecasting and technology would all be helped by a good understanding of maths. Other careers include social science research, any kind of statistical work, economics and business studies, work with computers, and of course many jobs in industry.

Preparing for GCSE

The GCSE syllabus is much more open-ended and manageable for the ordinary child than was O level. There is much less learning of boring theorems by rote and much more course work and an emphasis on investigational work.

This is a new aspect of maths for many of us and so I will explain briefly here what this means.

- It is about finding pattern in maths.
- It is about exploring mathematical thinking for ourselves.
- It is getting involved in maths at our own level and coming up with creative answers, not just the 'right' answer.

Here is just one example that I often use with adults. It might help you to see the point in this new way of teaching maths and if you want to follow this up, ask at school, or borrow Lorraine Mottershead's books (see page 174) from your local library.

Think of a stack of cans of beans in the supermarket. They might be stacked like this.

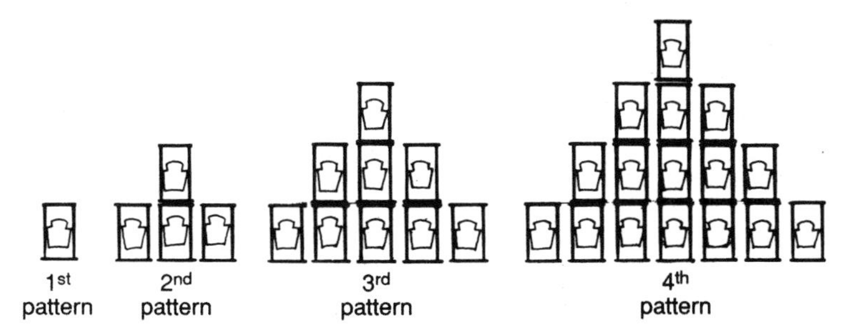

Can you see what the next pattern in the sequence would be? You might want to make it with lego bricks or by drawing it.

Now, if you look at the pattern that is developing, you may be able to see some number patterns in it.

- The rows across show the odd numbers (1, 3, 5, 7, etc.).

- The total number of cans in this pattern is 1 in the first pattern, then 4, 9, 16, 29, etc.

Do you recognise that pattern? They are the square numbers (1 squared is 1 (1×1), 2 squared is 4 ($2 \times 2 = 4$), $3 \times 3 = 9$, $4 \times 4 = 16$, etc.).

It is this playing around with numbers and *getting a feel for the mathematical thinking* involved that characterises GCSE.

If your child was doing the cans of beans investigation, she might be asked to think of how she could predict what the seventh pattern would look like . . . and then the thirty-seventh. You could think about that with her.

So now can you predict what the hundredth pattern would be like? How many bean cans in total?

Panic has probably set in by now! If you are really stuck, skip this, or come back to it with your child and work on it together.

You might have worked out that the seventh one would be 7 squared ($7 \times 7 = 49$) and the hundredth one would be 100 squared.

So how did you work that out?

The first one was 1 squared, the second one was 2 squared the third one was 3 squared and so on.

So, for any position in the pattern (i.e. second, third, etc.) we can say that it is the position in the sequence, squared. So for any size pile of cans I can think of, the total number of cans is the number along the bottom squared.

When I did maths with the Open University, we thought of the number of the position in a 'think cloud' – maybe it is 42. The total number of cans I would need for that forty-second pattern is the number in my think cloud, squared (or multiplied by itself), so it would be 42×42 cans.

Your child at school might call it 'position squared', or he might call it x squared. I like 'think clouds' because that way I don't have to use those dreadful xs.

OK so far? If you just read it through it might not be OK! Go

The think cloud

back and *do it*! Maths, contrary to popular belief, is not a spectator sport. You have to actually *do it*. Don't worry if you can't. Your child probably will be able to.

Now, the next point is, why is it 'squared'?

When I asked Sarah, aged seven, at a school that question, she went away and thought about it for a bit and then said 'I know, if you move that bit of the stack of beans around, you end up with a square'.

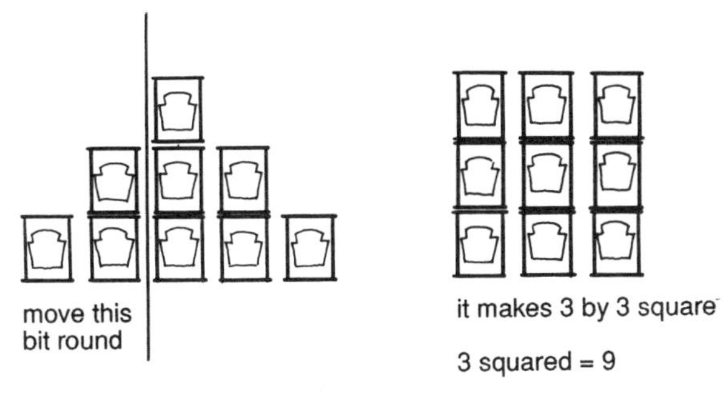

- Lots of adults can't see that at first! But this way of teaching maths is increasingly common in schools.
- Children are learning to think for themselves, not just learning what the teacher tells them.
- This way of learning tends to create mathematical thinkers.
- That's fun and challenging for the teacher, but sometimes a bit mind-blowing for us as parents.

I hope that as you read this book you have seen some of the

deeper meaning of maths. I hope that you will be able to see the maths that you do anyway in your daily life. I also hope that it might have taken some of the fear out of using maths. Losing that terror and fear of maths was a part of my own journey towards my delight in finding it absorbing and intriguing. I hope this book might have helped you on the same journey and given you confidence as you help your child.

> 'We have had the idea in mathematics that the child is an empty vessel and you pour in mathematics. Often we have thought of the child as a rather leaky vessel and mathematics has flowed out as well as in! But if we are to be more successful in helping children to learn mathematics, we need to see them as people who think about their experiences and we need to build on (these experiences).'
>
> (Hilary Shuard, in a BBC Horizon documentary 'Twice five plus the wings of a bird', 1988)

Appendix – resources

A recipe for play dough

3 lb plain flour
1 lb salt
2 tablespoons of any kind of cooking oil
about ¾ pint of warm water
some food colouring

Put the flour and salt in a bowl and mix. Add the oil and then put in the water a little at a time – don't let it get too sticky.

Divide the mixture into balls and give one to your child. Put a few drops of food colouring on the ball and they can knead the dough until the colour is thoroughly mixed in.

This mixture keeps well if you put it back in the plastic bag and seal it after use.

Further reading

If you want to read more about the ways in which children learn maths there are a number of books available. Many are listed in the individual chapters, but here are a few more.

How Children Learn Mathematics by Pamela Liebeck, Pelican Books. This is an easy to read account of the ways in which children learn maths. I use this book a great deal with students learning to be teachers.

Children's Minds by Margaret Donaldson, Fontana. This book has been of enormous influence in education as it discusses research that shows how much young children are capable of when things are presented to them in a way that they can understand.

Mathematics with Reason by Sue Atkinson, Hodder and Stoughton. This is a book of stories that I edited about the ways in which teachers have tried to teach maths to make children better mathematicians.

The Cockroft Report, HMSO Books. This widely-read report is available in libraries. The chapter on primary schools is particularly interesting. It covers secondary maths as well.

Other helpful books for younger children

Seven dizzy dragons, by Sue Atkinson, Sharon Harrison and Lynne McClure, Cambridge University Press.

Ten Beads Tall by Pam Adams, Child's Play. ISBN 0 85953 240 0.

Counting Kids by Annie Kubler, Child's Play. ISBN 0 85942 241 0.

Teach maths from stories by Sue Atkinson, Sharon Harrison and Lynne McClure, Cambridge University Press.

Useful addresses

The BEAM Project, Barnsbury Complex, Offord Road, London, N1 1QH. Write to it for a catalogue of its very good books or for details about courses on maths.

Tarquin Publications, Stradbroke, Diss, Norfolk, IP21 5JP. Send for the catalogue: it contains the most unusual, enjoyable and interesting maths publications available.

IMPACT, School of Education, University of North London, 1 Prince of Wales Road, London, NW5 3LB (IMPACT books are now published by Scholastic and are available in High Street shops.).

Galt Educational (for toys, games, Mottik, etc.) Brookfield Road, Cheadle, Cheshire, SK8 2PN (0161 428 8511).

Early Learning Centre (for toys and games) You can get a catalogue from them at South Marston, Swindon, SN3 4TJ (01793 832832), or look up your local centre in your telephone book.

Campaign for State Education (CASE) produces interesting publications including a bimonthly magazine 'Parents and schools' and other publications useful for parents, such as *Somebody ought to do something*. The main aim of CASE is to get the very best possible education for every child. It is not linked to any political party and arranges national and local conferences to get vital educational issues discussed. CASE, 158 Durham Road, London, SW20 0DG (0181 944 8206).

National Confederation of Parent-Teacher Associations will give advice on a wide range of topics. 2 Ebbsfleet Estate, Stonebridge Road, Gravesend, Kent, DA11 9DZ (01474 560618).

Appendix – the National Curriculum of England and Wales and the Scottish 5–14 Curriculum

The National Curriculum is slightly different in each part of the British Isles, but each is similar enough to the others to talk about some general points.

- Each curriculum is designed to give your child a 'balanced diet' of maths. This means that over a term or a month, your child will look at many different aspects of maths. She will, for example, do practical work, maths investigations, learn about shape and space, number, and use a computer data base.
- There is a focus on the child being able to use maths in everyday life. So he must learn what measuring is for and how to use it, not just how to measure lines printed in a maths text book. This is called 'problem solving' or 'using and applying maths' and it is a very important aspect of the maths your child is learning.

ATs (Attainment Targets)

ATs are the attainment targets. These are the major sections of the curriculum (such as 'handling data', or 'number'). The statements of attainment (below) are the little detailed bits of these attainment targets. There are five attainment targets in maths at the moment in the English and Welsh curriculum – but that is just about to be changed. We will know by September 1994 how many ATs there will be and we are told that this will then stay the same for five years.

SoAs (Statements of Attainment)

SoAs are specific targets in each subject that give details of the knowledge and skills that your child should learn and be tested on by their teachers. For example, in algebra and pattern, children at about seven 'should be able to explore number patterns'. In shape and space a seven-year-old 'should be able to use mathematical terms to describe common 2-D shapes and 3-D shapes'.

The five English and Welsh ATs (in 1994) are:

1. using and applying maths;
2. number and measures;
3. algebra;
4. shape and space;
5. handling data.

In Scotland there are four ATs:

1. problem solving and enquiry;
2. information handling;
3. number, money and measures;
4. shape, position and movement.

Testing and your child

If you are lucky enough to have a child at school in Scotland, official testing will be spread across a child's whole school life and the testing will take place when the child (or more likely, a small group of children) is ready for it. This links in very sensibly with the testing that teachers have always done and, of course, continue to do now. Do remember that you may, if you wish, withdraw your child from National Curriculum testing. Many parents do.

Wherever children are at school, they will be assessed in some way by the teacher at the end of a unit of work, or at the end of the term or year. Most of this assessment does not take the form of a written test, but is something that the teacher does as he works with the child. He observes the children working and listens carefully to what they say and makes notes for himself or for each child's folder which will be passed on to the next teacher.

Actually, teachers are assessing your child all the time, not just at special times. As they plan work they are thinking 'this is going to be too hard for Sandy and Mike, I'm going to have to do something different for them'.

As they sit with the children around them they are noting who is happy with an idea and who still looks puzzled. They learn to distinguish between the child who is not putting up her hand because she doesn't know and the child who is quiet because she is shy. In fact, all the time your child is at school she is being assessed. It has always been like that, but what is new is that teachers are now required by law to do certain tests.

SATS and your child at seven

The Standard Assessment Tests (SATs) are changing all the time as it has been recognised that if teachers of seven-year-olds spend too much time testing, very little learning can take place in the class. The trouble is that a short written test doesn't tell you very much about a seven-year-old!

Your child's teacher will give your child plenty of practice with the layout of the SATs and will read it through with the children before they have to complete it.

For example, at level 2 there was a question that involved knowing some number patterns. The children had to work out what went in each box and write it in.

1, 2, 3, □, 5, 6.
2, 1, 2, 1, □, 1
2, 4, □, 8, 10, 12.
10, 9, 8, □, 6, 5.

$4 + 2 = \square$
$5 - \square = 4$
$\square + 7 = 8$
$\square + 3 = 5$

Your child will usually work in a small group with the teacher. Children are allowed to use counters as part of the test except for the questions at levels 2 and 3 that are about their knowledge of number bonds. For these questions they have to work mentally.

There have been many changes to the tests but the emphasis of the tests will be on number.

Teacher assessment

This assessment is of a more informal type and often your child will not know that she is being assessed. It is done during normal classroom activity and goes on most of the time. When your child does a piece of work, or answers at mental maths time, the teacher is assessing what that child needs to do next.

These assessments will make up the report that the teacher will make on your child at the end of the year.

The Key Stages and the average levels of pupils

What is crucial for parents and teachers is not to let the child feel anxious about his tests. That will only make him fear maths and perhaps get nightmares and perform less well in the tests than if they were just treated as a part of everyday classroom life. Like any test, they can only give you a small amount of information about that child on that particular day with that particular test.

Most children will achieve at an average level – obviously. The tests are based on what most children can do at a particular age. So most seven-year-olds will achieve level 2, or level A in Scotland. If your child achieves below average it may only mean that she was not coping well that day, or that she misunderstood the question, or that the test wasn't a very good one, or that your child needs a bit more time to develop in some areas.

Similarly, a child achieving above the average is not necessary a genius!

One seven-year-old test on probability that I watched a colleague do with her class gave almost all the children an above average mark! We thought that this was partly because the test was quite easy but also because it hadn't been all that long since the children had done that work in class, so they had all remembered it. Results might have been quite different in different circumstances.

The really crucial thing to remember is that tests are not necessarily the way to assess ability successfully. What your child can do, what he remembers, the skills and concepts she has mastered, are all very complex. The words by which your child's teacher explains what your child has done may well give you more information than some test result.

Tests for eleven-year-olds

These tests will be taken near the end of Year 6 (P7 in Scotland), usually when the child is about eleven. This is the end of Key Stage 2.

What will the grades mean?

The child will be given a grade indicating the level that she has reached in each subject. The grades at eleven in maths will vary enormously. It is widely accepted that by eleven there is an ability gap of seven years between the brightest children and those with learning difficulties.

So some children are able to learn like 14-year-olds at this stage, but others are still like seven-year-olds.

Most children at 11 will achieve at level 4 (D in Scotland) in maths. Levels 2 and 3 would be below average and level 5 would be above average. If a child attained level 6, that would be exceptional for an eleven year old.

What will the tests involve?

The tests keep changing, but after a few years they will probably settle down. Here is a question from the 1993 pilot tests for 11-year-olds. You can expect fairly similar things to be in your child's test.

> There are six mini-rolls in each pack.
> They buy 25 packs.
> How many mini-rolls did they buy? ☐

There was also a question on reflective symmetry, place value and shape. In other words, all the questions will be about the maths that your child is doing anyway.

Tests for seven-year-olds

These tests are well established now and teachers are used to keeping children calm and helping them to do their best and not feel anxious. At first the tests were very long and involved, but now that they are simplified they do not disrupt the class as much.

The questions are very much what you would expect. In the tests in 1993 there was some simple addition and subtraction at level 1.

$4 + 5 =$
$6 + 2 =$
$8 - 3 =$
$7 - 5 =$

This table shows when children are tested in England and Wales, and the probable levels of ability that will be achieved.

Key Stage	testing at	level of the majority of pupils
1	7	1–3
2	11	2–5
3	14	5–7
4	16 (GCSE)	4–10

In England and Wales, although you will get a report on your child each year, the reporting of the levels is only at the end of the Key Stages.

In Scotland

Level A	should be attainable during P1–P3 by most pupils.
Level B	should be attainable by some in P3 and most by P4.
Level C	should be attainable during P4–6 by most pupils.
Level D	should be attainable by some in P5 or 6, and by most in P7.
Level E	should be attainable by some in P7–S1 and by most in S2.

Answers to the button puzzles

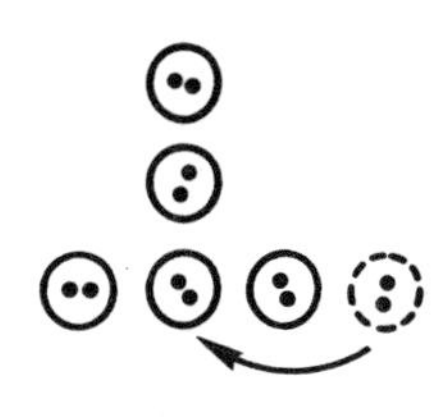

1 Put the one at the far right on top of the one that is in both rows.

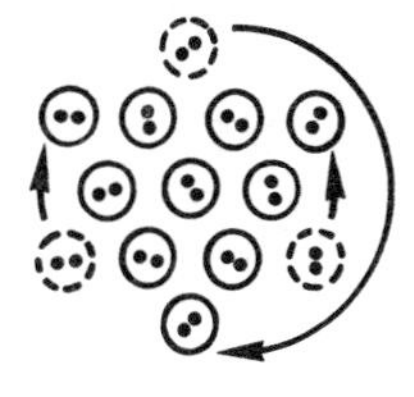

2 Move the top counter down, and the two at the edge up.

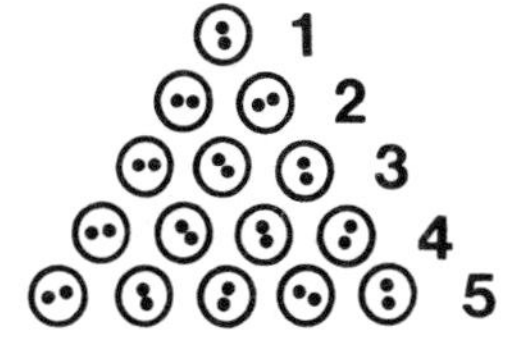

3 The next pattern is . . .
You add a row at the bottom each time.

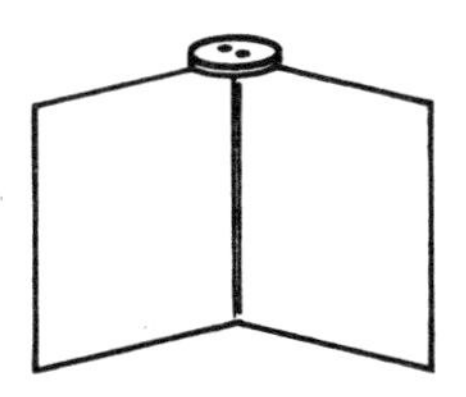

4 It's a trick! Fold the paper in half and balance the button on the fold.

Index